Lecture Notes in Mathematics

Edited by A. Dold and B. Eckmann

Subseries: Mathematatisches Institut der Universität und
Max-Planck-Institut für Mathematik, Bonn – vol. 4
Adviser: F. Hirzebruch

1102

Andrei Verona

Stratified Mappings – Structure and Triangulability

Springer-Verlag
Berlin Heidelberg New York Tokyo 1984

Author

Andrei Verona
Department of Mathematics and Computer Science
California State University
Los Angeles, CA 90032, USA

AMS Subject Classification (1980): 57R05, 58A35; 32B25, 32C42, 57R35, 57R45, 57S15, 58C25, 58C27

ISBN 3-540-13898-6 Springer-Verlag Berlin Heidelberg New York Tokyo
ISBN 0-387-13898-6 Springer-Verlag New York Heidelberg Berlin Tokyo

Printing and binding: Beltz Offsetdruck, Hemsbach/Bergstr.
2146/3140-543210

INTRODUCTION

For several reasons, most of them stemming from algebraic topology, it is important to know whether a topological space, or more generally a continuous map, is triangulable or not. Cairns [Ca] proved the triangulability of smooth manifolds; another proof, also providing a uniqueness result, is due to J. H. C. Whitehead [Wh]. First attempts to prove the triangulability of algebraic sets are due van der Waerden [W], Lefschetz [Le], Koopman and Brown [K-B] and Lefschetz and Whitehead [L-W]. Rigorous proofs, in the more general case of semianalytic sets, were given by Lojasiewicz [Lo] and Giesecke [Gi]. Later, Hironaka [Hi_1] and Hardt [Ha_2] proved the triangulability of subanalytic sets. The most general spaces known to be triangulable are the stratified sets introduced by Thom [T_2] and the abstract stratifications introduced by Mather [Ma_1] (Mather's notion is slightly different from Thom's one, but it is more or less clear that the two classes of spaces coincide, at least in the compact case); they include all the spaces mentioned above and also the orbit spaces of smooth actions of compact Lie groups. Their triangulability was proved by several authors (Goresky [G_1], Johnson [J_2], Kato [Ka] and Verona [Ve_3] to mention only the published proofs. A more detailed discussion of these proofs and of others can be found in the introduction of Johnson's paper or at the end of Section 7 of the present work). The more difficult problem of the triangulability of mappings was considered by much fewer authors: Putz [P] proved the triangulability of smooth submersions, Hardt [Ha_2] proved the triangulability of some, very special, subanalytic maps and I proved in [Ve_3] the triangulability of certain stratified maps. In [T_1] Thom considered the problem of the triangulability of smooth maps and (implicitely) conjectured that "almost all" smooth mappings are triangulable. It is the aim of this paper to prove this conjecture. To be precise, we shall prove

<u>Theorem</u>. Let M and N be smooth manifolds without boundary. Then any proper, topologically stable smooth map from M to N is triangulable.

Since the set of proper and topologically stable smooth maps from M to N is dense in the set of all proper smooth maps from M to N (Thom-Mather theorem*) we obtain a positive answer to the above mentioned conjecture.

As a matter of fact, we prove a more general result concerning the triangulability of certain stratified mappings (Theorem 8.9) which implies the theorem stated above and also the following result (first proved by Hardt [Ha_2]): any proper light subanalytic map is triangulable (light means that the preimage of a discrete set is discrete). Our Theorem 8.9 is not as general as one would expect it. It applies only to proper and nice abstract Thom mappings (nice means that the mapping is finite to one when restricted to a certain subspace). It is natural to conjecture that the theorem is true for any proper abstract Thom mapping. A positive answer would solve another conjecture of Thom [T_1] : any proper Thom mapping (in Thom's terminology "application sans eclatement") is triangulable. The main difficulty in proving this more general version of Theorem 8.9 is explained in Section 8.15.3.

Since we are dealing with stratified spaces as introduced by Mather in [Ma_1] and since these lecture notes have never been published, I thought it would be useful to collect in a first part of the present work (Chapters 1, 2, and 3) the main results of the theory. Some of the proofs presented here are new and simpler than the original ones. For technical reasons, we are obliged to work with certain manifolds with corners, called here manifolds with faces. The necessary facts concerning them are presented in Chapter 4. In Chapter 5, we extend the theory of abstract strati-

---

*) This result was conjectured by Thom [T_1]. In [T_3] and [T_2] he outlined a possible proof. Later Mather ([Ma_2] and [Ma_3]) filled in the details, made it rigorous and, by slightly changing some of Thom's concepts, cleaned up many technical points. Another proof (only slightly different) is presented in [Gib].

fications and abstract Thom mappings to the case when the strata are allowed to be manifolds with faces; most of the proofs are copies of the proofs presented in the first three chapters and so they are omitted. In Chapter 6, we prove some theorems concerning the structure of abstract stratifications and of abstract Thom mappings. In some sense, they can be viewed as a kind of "resolution of the singularities" in the C^∞-case. For example, Theorem 6.5 can be interpreted as saying that any abstract stratification of finite depth can be obtained from a manifold with faces by making certain identifications on the faces. Chapter 8 contains the main results of the paper (they were mentioned above). In an appendix, I have collected some facts from PL-topology which are needed in Chapters 7 and 8.

Acknowledgement. For their support and/or hospitality during the preparation of this work, I wish to express my gratitude to the Institut des Hautes Etudes Scientifiques in Bures-sur-Yvette, the Alexander von Humboldt Foundation, the Sonderforschungsbereich Theoretische Mathematik of the University of Bonn, the National Science Foundation (Grant MCS-8108814 (A01), the Institute for Advanced Study in Princeton and, last but not least, the Max-Planck-Institut für Mathematik in Bonn. Especially, I wish to thank Professor F. Hirzebruch, whose confidence and understanding during the preparation of this paper were of great help.

TABLE OF CONTENTS

0. NOTATION AND CONVENTION

0.1. A topological space is called <u>nice</u> if it is Hausdorff, locally compact, paracompact and with a countable basis for its topology.

0.2. If A is a topological space and $X \subset A$, then $cl_A(X)$ (resp. $int_A(X)$) denotes the closure (resp. interior) of X in A.

0.3. If A and B are topological spaces, then $A \sqcup B$ denotes their disjoint union.

0.4. For any set A, 1_A (or id_A) denotes the identity map of A.

0.5. Smooth means always differentiable of class C^∞.

0.6. The connected components of a smooth manifold may have different dimensions. Given a smooth manifold M we denote by TM its tangent bundle. If $x \in M$, TM_x denotes the tangent space of M at x. If N is another smooth manifold and $f : M \to N$ is smooth, $df : TM \to TN$ denotes the differential of f; if $x \in M$, then $df_x : TM_x \to TN_{f(x)}$ denotes the restriction of df. The smooth map f is called <u>submersive</u> if df_x is surjective for any $x \in M$.

0.7. Let A be a topological space and let X, Y and Z be subsets of A. Let f and g be maps defined on X and Y respectively. We say that f <u>equals</u> g <u>near</u> Z (denoted $f = g$ near Z) if there exists a neighborhood U of Z such that $f|X \cap U = g|Y \cap U$. The same terminology is also used in other similar situations (for example $X = Y$ near Z means that there exists a neighborhood U of Z such that $X \cap U = Y \cap U$).

0.8. R denotes the field of real numbers ; $R_+ = \{r \in R \,;\, r \geq 0\}$ and $R_+^* = \{r \in R \,;\, r > 0\}$.

1. ABSTRACT STRATIFICATIONS

1.1. Let A be a nice topological space and let $X \subset A$ be a locally closed subset. Let T_X be an open neighborhood of X in A, $\pi_X : T_X \longrightarrow X$ be a continuous retraction (i.e., $\pi_X | T_X = 1_X$) and $\rho_X : T_X \longrightarrow R_+$ be continuous and such that $\rho_X^{-1}(0) = X$. Given $\varepsilon, \delta : X \longrightarrow R$ we shall use the following notation $\varepsilon < \delta$ if $\varepsilon(x) < \delta(x)$ for any $x \in X$; if $\varepsilon < \delta$ set

$$X \times (\varepsilon, \delta) = \{(x, t) \in X \times R;\ \varepsilon(x) < t < \delta(x)\}$$

$$X \times \{\varepsilon\} = \{(x, t) \in X \times R;\ t = \varepsilon(x)\}$$

$$X \times [\varepsilon, \delta) = (X \times (\varepsilon, \delta)) \cup (X \times \{\varepsilon\})$$

$$X \times (\varepsilon, \delta] = (X \times (\varepsilon, \delta)) \cup (X \times \{\delta\})$$

$$X \times [\varepsilon, \delta] = (X \times [\varepsilon, \delta)) \cup (X \times \{\delta\})$$

$$T_X^\varepsilon = \{a \in T_X;\ \rho_X(a) < \varepsilon(\pi_X(a))\}$$

$$S_X^\varepsilon = \{a \in T_X;\ \rho_X(a) = \varepsilon(\pi_X(a))\}\ .$$

In such an expression a real number will be considered as the corresponding constant function on X. From now on, unless the contrary is specified, ε (or $\varepsilon_1, \varepsilon_2, \ldots, \delta, \delta_1, \ldots$) will be a continuous function taking values in R_+^*, its domain being determined by the context; if in addition the domain is a smooth manifold, then ε will be assumed to be smooth. When no confusion can arise we shall denote the restrictions of π_X and ρ_X to T_X^ε by the same symbols; otherwise we shall use the notation π_X^ε and ρ_X^ε respectively.

Let T_X, π_X and ρ_X be as above. One can verify that, possibly after shrinking T_X, the following assertions are true:

(1.1.1) if $X \subset U \subset A$ with U open, then $T_X^\varepsilon \subset U$ for some ε;

(1.1.2) $(\pi_X^\varepsilon, \rho_X^\varepsilon) : T_X^\varepsilon \longrightarrow X \times [0, \varepsilon)$ is proper for some ε.

As an immediate consequence

(1.1.3) given a compact subset $K \subset X$ and a neighborhood U of K in A, there exists $\varepsilon \in R_+^*$ and a neighborhood V of K in X such that $\pi_X^{-1}(V) \cap T_X^\varepsilon \subset U$.

Let T'_X, π'_X and ρ'_X have the same properties as T_X, π_X and ρ_X. The triples (T_X, π_X, ρ_X) and (T'_X, π'_X, ρ'_X) are called equivalent if $(T_X, \pi_X, \rho_X) = (T'_X, \pi'_X, \rho'_X)$ near X, that is, there exists a neighborhood U of X in A such that $T_X \cap U = T'_X \cap U$, $\pi_X | T_X \cap U = \pi'_X | T_X \cap U$ and $\rho_X | T_X \cap U = \rho'_X | T_X \cap U$. By definition, a <u>tube</u> of X in A is an equivalence class of triples (T_X, π_X, ρ_X). In order to simplify the notation, we shall not distinguish between a tube and a triple which represents it; however we shall consider only triples which verify (1.1.1) and (1.1.2) (and therefore (1.1.3) too).

Let now $Y \subset A$ be another locally closed subset such that $X \subset c\ell_A(Y)$ and $X \cap Y = \phi$ (in this situation we shall write $X < Y$; as usual $X \leq Y$ means that $X < Y$ or $X = Y$). Let $\tau_X = (T_X, \pi_X, \rho_X)$ and $\tau_Y = \{T_Y, \pi_Y, \rho_Y\}$ be tubes of X and Y respectively. We shall consider "<u>control conditions</u>" of the form

(1.1.4) there exist ε and δ such that $a \in T_X^\varepsilon \cap T_Y^\delta$ implies $\pi_Y(a) \in T_X$ and $\pi_X(\pi_Y(a)) = \pi_X(a)$;

(1.1.5) there exist ε and δ such that $a \in T_X^\varepsilon \cap T_Y^\delta$ implies $\pi_Y(a) \in T_X$ and $\rho_X(\pi_Y(a)) = \rho_X(a)$.

Next let $f : B \longrightarrow A$ be a continuous map, where B is another nice topological space. Let $Y \subset B$ and $X \subset A$ be locally closed subsets such that $f(Y) \subset X$. Given tubes τ_X and τ_Y of X in A and Y in B respectively, we shall also consider control conditions of the form

(1.1.6) there exists δ such that $f(T_Y^\delta) \subset T_X$ and

$f(\pi_Y(b)) = \pi_X(f(b))$, $b \in T_Y^\delta$;

(1.1.7) there exists δ such that $f(T_Y^\delta) \subset T_X$ and

$\rho_Y(b) = \rho_X(f(b))$, $b \in T_Y^\delta$.

1.2.1. A <u>weak abstract stratification</u> (w.a.s.) $\underline{\underline{A}}$ consists of (i) a nice topological space A; (ii) a locally finite family $\mathcal{A}$ of locally closed subsets of A (called <u>strata</u>) such that A is the disjoint union of the strata; (iii) a family of tubes of the strata, $\{\tau_X;\ X \in \mathcal{A}\}$. The strata and their tubes must satisfy the following four axioms:

(1.2.1.1) if $X, Y \in \mathcal{A}$ and $X \cap c\ell_A(Y) \neq \phi$, then $X \leq Y$;

(1.2.1.2) each stratum is a smooth manifold (without boundary) in the induced topology;

(1.2.1.3) for any $X \in \mathcal{A}$ there exists ε_X such that for any stratum $Y \neq X$ of $\underline{\underline{A}}$, $T_X \cap Y \neq \phi$ implies that $X < Y$ and $(\pi_X, \rho_X)|T_X^{\varepsilon_X} \cap Y : T_X^{\varepsilon_X} \cap Y \longrightarrow X \times (0, \varepsilon_X)$ is smooth and submersive (in particular $\dim(X) < \dim(Y)$);

(1.2.1.4) for any strata $X < Y$, the tubes τ_X and τ_Y satisfy (1.1.4) with $\varepsilon = \varepsilon_X$ and $\delta = \varepsilon_Y$.

If in addition

(1.2.1.5) for any strata $X < Y$, the tubes τ_X and τ_Y satisfy (1.1.5) with $\varepsilon = \varepsilon_X$ and $\delta = \varepsilon_Y$,

then $\underline{\underline{A}}$ is called an <u>abstract stratification</u> (a.s.).

Sometimes we shall say that $\underline{\underline{A}}$ is a <u>(w.)a.s. structure</u> on A.

Let $\underline{\underline{A}}$ be a w.a.s. Since A is a normal space, we can assume without loss of generality that

(1.2.1.6) let $X, Y \in A$; then $T_X^{\varepsilon_X} \cap T_Y^{\varepsilon_Y} \neq \phi$ if and only if $X \leq Y$ or $Y < X$.

<u>Remark</u>. Let $\underline{\underline{A}}$ (resp. $\underline{\underline{A}}'$) be a w.a.s. with tubes $\tau_X = \{T_X, \pi_X, \rho_X\}$, $X \in \mathcal{A}$ (resp. $\tau_{X'} = \{T'_{X'}, \pi'_{X'}, \rho'_{X'}\}$, $X' \in \mathcal{A}'$). It is important to notice that, in view of our earlier convention on tubes, $\underline{\underline{A}} = \underline{\underline{A}}'$ if and only if $A = A'$, $\mathcal{A} = \mathcal{A}'$ and $(T_X, \pi_X, \rho_X) = (T'_X, \pi'_X, \rho'_X)$ near X, for any $X \in \mathcal{A}$. Thus it may happen that, for some $X \in \mathcal{A}$, $T_X \neq T'_X$ or $\pi_X \neq \pi'_X$ or $\rho_X \neq \rho'_X$, but $T_X = T'_X$ near X, $\pi_X = \pi'_X$ near X and $\rho_X = \rho'_X$ near X.

1.2.2. Let $\underline{\underline{A}}$ be a w.a.s. and $X \in \mathcal{A}$ be a stratum. Define the <u>depth of X in $\underline{\underline{A}}$</u> to be the integer

$$\mathrm{depth}_{\underline{\underline{A}}}(X) = \sup\{n;\ \text{there exist strata } X = X_0 < X_1 < \ldots < X_n\} .$$

Next define the <u>depth</u> of $\underline{\underline{A}}$ and the <u>dimension</u> of $\underline{\underline{A}}$ to be

$$\mathrm{depth}(\underline{\underline{A}}) = \sup\{\mathrm{depth}_{\underline{\underline{A}}}(X);\ X \in \mathcal{A}\}$$

and

$$\dim(\underline{\underline{A}}) = \sup\ \{\dim(X);\ X \in \mathcal{A}\} .$$

One can prove that $\dim(\underline{\underline{A}})$ is the topological dimension of A and thus is a topological invariant. Although $\mathrm{depth}(\underline{\underline{A}})$ is obviously not a topological invariant, it is clear that $\dim(\underline{\underline{A}}) < \infty$ implies that $\mathrm{depth}(\underline{\underline{A}}) < \infty$. Notice also that $\mathrm{depth}(\underline{\underline{A}}) = 0$ if and only if all the strata are open and closed in A, i.e., if and only if A is a smooth manifold and each stratum $X \in \mathcal{A}$ is the union of some connected components of A.

1.2.3. Let $\underline{\underline{A}}$ be a (w.)a.s. and $U \subset A$ be a locally closed subset. Set $\mathcal{A}|U = \{X \cap U;\ X \in \mathcal{A}$ and $X \cap U \neq \phi\}$. Suppose that for any $X \cap U \in \mathcal{A}|U$ there is given $T_{X\cap U} \subset U \cap \pi_X^{-1}(X \cap U)$ containing $X \cap U$ and set $\pi_{X\cap U} = \pi_X | T_{X\cap U} : T_{X\cap U} \longrightarrow X \cap U$, $\rho_{X\cap U} = \rho_X | T_{X\cap U} : T_{X\cap U} \longrightarrow R_+$ and $\tau_{X\cap U} = \{T_{X\cap U}, \pi_{X\cap U}, \rho_{X\cap U}\}$. If

(1.2.3.1) each $T_{X \cap U}$ is open in U;

(1.2.3.2) $X \cap U$, $Y \cap U \in \underline{\underline{A}}|U$ and $(X \cap U) \cap c\ell_U(Y \cap U) \neq \phi$ imply that $X \cap U \subset c\ell_U(Y \cap U)$;

(1.2.3.3) any $X \cap U \in \underline{\underline{A}}|U$ is a smooth submanifold of X ;

(1.2.3.4) any $X \cap U \in \underline{\underline{A}}|U$ verifies (1.2.1.3) (with X and Y replaced by $X \cap U$ and $Y \cap U$ respectively),

then $\underline{\underline{A}}|U$ and $\{\tau_{X \cap U}; X \cap U \in \underline{\underline{A}}|U\}$ determine a (w.)a.s. structure on U, called the restriction of $\underline{\underline{A}}$ to U and denoted $\underline{\underline{A}}|U$. From (1.2.3.1) and the inclusions $X \cap U \subset T_{X \cap U} \subset U \cap \pi_X^{-1}(X \cap U)$ it follows that $U \cap \pi_X^{-1}(X \cap U)$ is open in U near $X \cap U$ and thus

$$T_{X \cap U} = U \cap \pi_X^{-1}(X \cap U) \quad \text{near} \quad X \cap U .$$

As a consequence $\underline{\underline{A}}|U$ (if it exists) is completely determined by $\underline{\underline{A}}$ and U.

It is also important to notice that if U is locally closed in A, V is locally closed in U and $\underline{\underline{A}}|U$ exists, then $(\underline{\underline{A}}|U)|V$ exists if and only if $\underline{\underline{A}}|V$ exists and if they exist then

$$\underline{\underline{A}}|V = (\underline{\underline{A}}|U)|V \quad . \tag{1.2.3.5}$$

The simplest examples of subsets $U \subset A$ for which $\underline{\underline{A}}|U$ exists are provided by locally closed subsets which are union of strata or by open subsets. Other examples are given in 2.9 and 2.12.

1.2.4. Let $\underline{\underline{A}}$ and $\underline{\underline{B}}$ be w.a.s.'s. A continuous map $f : B \longrightarrow A$ is called a weak morphism, denoted $f : \underline{\underline{B}} \dashrightarrow \underline{\underline{A}}$, if for any $Y \in \mathcal{B}$ there exists $X \in \mathcal{A}$ such that $f(Y) \subset X$, $f|Y : Y \longrightarrow X$ is smooth and the tubes τ_X and τ_Y satisfy (1.1.6). If in addition τ_X and τ_Y satisfy (1.1.7), then f is called a morphism and is denoted $f : \underline{\underline{B}} \longrightarrow \underline{\underline{A}}$. The differential of $f|Y$ will be denoted simply df.

It is useful to notice that if $f : B \longrightarrow A$ is not known to be continuous, but it verifies all the other conditions in the definition of a morphism, then it is continuous, hence it is a morphism. This follows from (1.1.3), (1.1.6), (1.1.7) and the continuity of f on the strata. A similar assertion is true for weak morphisms if $\operatorname{depth}(\underline{\underline{A}}) = 0$.

1.2.5. A <u>(weak) isomorphism</u> is a (weak) morphism whose inverse (as a map) exists and is a (weak) morphism. Clearly a (weak) morphism is a (weak) isomorphism if and only if it is a homeomorphism sending strata diffeomorphically onto strata.

For example, given two w.a.s.'s $\underline{\underline{A}}$ and $\underline{\underline{A}}'$ with $A = A'$, 1_A is an isomorphism if and only if $\underline{\underline{A}}$ and $\underline{\underline{A}}'$ are equal. If $A = A'$ and 1_A is a weak isomorphism, we shall call $\underline{\underline{A}}$ and $\underline{\underline{A}}'$ <u>weakly equal</u>.

1.2.6. $f : \underline{\underline{B}} \longrightarrow \underline{\underline{A}}$ is called <u>submersive</u> if for any $Y \in \mathcal{B}$ and $X \in \mathcal{A}$ with $f(Y) \subset X$, $f|Y : Y \longrightarrow X$ is submersive.

1.2.7. A smooth manifold M, if not connected, can be endowed with different a.s. structures of depth zero. By $\underline{\underline{M}}$ we shall denote the unique a.s. structure on M for which $M \in \mathcal{M}$ (thus $\mathcal{M} = \{M\}$, $T_M = M$, $\pi_M = 1_M$ and $\rho_M = 0$).

1.2.8. Let $\underline{\underline{A}}$ be a w.a.s. and M be a smooth manifold. A map $f : A \longrightarrow M$ is called <u>controlled</u> (resp. a <u>controlled submersion</u>) if it is a weak morphism (resp. a submersive weak morphism) from $\underline{\underline{A}}$ to $\underline{\underline{M}}$. If $U \subset A$ is locally closed and $\underline{\underline{A}}|U$ exists, we shall denote by $C^\infty_{\underline{\underline{A}}}(U, M)$ the set of all controlled maps from $\underline{\underline{A}}|U$ to M. Clearly the family $\{C^\infty_{\underline{\underline{A}}}(U, M);\ U$ open in $A\}$ together with the obvious restrictions $C^\infty_{\underline{\underline{A}}}(U, M) \longrightarrow C^\infty_{\underline{\underline{A}}}(V, M)$ (for $V \subset U$) is a sheaf on A. If $M = R$ we shall denote $C^\infty_{\underline{\underline{A}}}(U, R)$ by $C^\infty_{\underline{\underline{A}}}(U)$ and the corresponding sheaf (of real algebras) by $C^\infty_{\underline{\underline{A}}}$. If $U \subset A$ is locally closed and $\underline{\underline{A}}|U$ exists, an element of $C^\infty_{\underline{\underline{A}}}(U)$ is called a <u>controlled function</u> on U.

1.2.9. Given two w.a.s.'s $\underline{\underline{A}}$ and $\underline{\underline{B}}$ we define their <u>product</u> $\underline{\underline{A}} \times \underline{\underline{B}}$ as follows. The underlying topological space of $\underline{\underline{A}} \times \underline{\underline{B}}$ is $A \times B$; the family of strata is $\mathcal{A} \times \mathcal{B} = \{X \times Y;\ X \in \mathcal{A},\ Y \in \mathcal{B}\}$. For $X \times Y \in \mathcal{A} \times \mathcal{B}$ set $T_{X\times Y} = T_X \times T_Y$, $\pi_{X\times Y}(a, b) = (\pi_X(a), \pi_Y(b))$, $\rho_{X\times Y}(a, b) = \rho_X(a) + \rho_Y(b)$ and $\tau_{X\times Y} = (T_{X\times Y}, \pi_{X\times Y}, \rho_{X\times Y})$. There is no problem in verifying that, in this way, we have defined a w.a.s. However it is important to notice that if $\underline{\underline{A}}$ and $\underline{\underline{B}}$ are a.s.'s then $\underline{\underline{A}} \times \underline{\underline{B}}$ is an a.s. if and only if $\mathrm{depth}(\underline{\underline{A}})\cdot\mathrm{depth}(\underline{\underline{B}}) = 0$. Indeed, if $\mathrm{depth}(\underline{\underline{A}})\cdot\mathrm{depth}(\underline{\underline{B}}) > 0$ there exist strata $X < X'$ of $\underline{\underline{A}}$ and $Y < Y'$ of $\underline{\underline{B}}$; then for $a \in T_X^{\varepsilon} \cap X'$ and $b \in T_Y^{\delta} \cap T_{Y'}^{\delta'}$, with ε, δ and δ' sufficiently small, we have $\pi_{Y'}(b) \in T_Y$ and

$$\rho_{X\times Y}(\pi_{X\times X'}(a, b)) = \rho_{X\times Y}(\pi_X(a), \pi_{Y'}(b)) =$$
$$= \rho_X(\pi_X(a)) + \rho_Y(\pi_{Y'}(b)) = \rho_Y(b)$$

while

$$\rho_{X\times Y}(a, b) = \rho_X(a) + \rho_Y(b) \neq \rho_Y(b)$$

and thus condition (1.2.1.5) cannot be satisfied. If $\mathrm{depth}(\underline{\underline{A}}) = 0$ or $\mathrm{depth}(\underline{\underline{B}}) = 0$, then it is easy to check that $\underline{\underline{A}} \times \underline{\underline{B}}$ is an a.s. as soon as $\underline{\underline{A}}$ and $\underline{\underline{B}}$ are a.s.'s.

If $\underline{\underline{A}}$ is a w.a.s., $\varepsilon, \delta : A \longrightarrow R$ are continuous and $\varepsilon < \delta$, then the w.a.s. $(\underline{\underline{A}} \times \underline{\underline{R}})|A \times (\varepsilon, \delta)$ will be denoted $\underline{\underline{A}} \times (\varepsilon, \delta)$.

1.3. LEMMA. Let $\underline{\underline{A}}$ be a w.a.s. Then for any covering of A by open subsets there exists a subordinate partition of unity consisting of controlled functions.

Proof. Since A is paracompact, the assertion is equivalent to the following local statement:

"Let $x \in A$ and V be a neighborhood of x in A. Then there exists $f \in C^{\infty}_{\underline{\underline{A}}}(A)$ such that $f(A) \subset [0, 1]$, $\mathrm{supp}(f) \subset V$ and $f^{-1}(1)$ is a neighborhood of x in A."

The statement being local, it suffices to consider w.a.s.'s of finite depth. If $\operatorname{depth}(\underline{\underline{A}}) = 0$, A is a smooth manifold and all smooth functions are controlled. The statement follows. Assume inductively that the statement is true for any w.a.s. of depth less than $\operatorname{depth}(\underline{\underline{A}})$ and let x and V be as above. Let $X \in \mathcal{A}$ be the stratum containing x and let $\varepsilon : X \longrightarrow R_+^*$ be such that all the conditions involved in the definition of a w.a.s. are verified on T_X^ε. Let also K be a compact neighborhood of x in X. Choose $g \in C^\infty(X)$ such that $g(X) \subset [0, 1]$, $\operatorname{supp}(g) \subset K$ and $g^{-1}(1)$ is a neighborhood of x in X. Define $g_1 \in C^\infty_{\underline{\underline{A}}}(T_X^\varepsilon)$ by $g_1 = g \circ \pi_X^\varepsilon$. Next consider $\underline{\underline{A}}|T_X^\varepsilon \setminus X$ and notice that $\operatorname{depth}(\underline{\underline{A}}|T_X^\varepsilon \setminus X) < \operatorname{depth}(\underline{\underline{A}})$. Therefore, by induction, there exists $h \in C^\infty_{\underline{\underline{A}}}(T_X^\varepsilon \setminus X)$ such that $h(T_X^\varepsilon \setminus X) \subset [0, 1]$, $\operatorname{supp}(h) \subset T_X^\delta$ for some $\delta < \varepsilon$ and $h(a) = 1$ for any $a \in T_X^{\delta/2} \setminus X$. Define $h_1 \in C^\infty_{\underline{\underline{A}}}(T_X^\varepsilon)$ by $h_1|T_X^\varepsilon \setminus X = h$ and $h_1(a) = 1$ if $a \in X$. In view of (1.1.3) we can choose δ and K such that $\pi_X^{-1}(K) \cap T_X^\delta \subset V$. Finally define $f \in C^\infty_{\underline{\underline{A}}}(A)$ by setting

$$f(a) = \begin{cases} g_1(a)h_1(a), & a \in T_X^\varepsilon \\ 0, & a \notin T_X^\delta \end{cases} .$$

It is obvious that f has the required properties. Q.E.D.

1.4. COROLLARY. The sheaf $C^\infty_{\underline{\underline{A}}}$ is fine.

Notes. The definition of abstract stratifications given above is due to Mather $[M_1]$ (he calls them abstract prestratifications). Lemma 1.3 appears first in $[V_1]$ and $[V_2]$; the proof given here is simpler.

2. CONTROLLED VECTOR FIELDS

2.1. A vector field on a w.a.s. $\underline{\underline{A}}$ is a collection $\xi = \{\xi(x) \in TX_x;\ x \in X,\ X \in \mathcal{A}\}$ such that for any $X \in \mathcal{A}$

(2.1.1) $X \ni x \longmapsto \xi(x)$ is a smooth vector field on X.

The vector field ξ is called <u>weakly controlled</u> if

(2.1.2) for any $X \in \mathcal{A}$ there exists $\varepsilon < \varepsilon_X$ (ε_X as in (1.2.1.3)) such that for any $Y \in \mathcal{A}$ with $X < Y$ and any $y \in T_X^\varepsilon \cap Y$, $d\pi_X \cdot \xi(y) = \xi(\pi_X(y))$.

If in addition, the notation being as above,

$$d\rho_X \cdot \xi(y) = 0 \quad , \tag{2.1.3}$$

then ξ is called a <u>controlled vector field</u> on $\underline{\underline{A}}$.

Given a subset $U \subset A$, set $\xi|U = \{\xi(x);\ x \in U\}$. Thus (2.1.1) means that, for any $X \in \mathcal{A}$, $\xi|X$ is a smooth vector field on X.

Suppose now that $U \subset A$ is locally closed and $\underline{\underline{A}}|U$ exists; a (weakly) controlled vector field on $\underline{\underline{A}}|U$ is called a (weakly) controlled vector field on U. The set of all weakly controlled (resp. controlled) vector fields on U is denoted $\mathcal{X}^w_{\underline{\underline{A}}}(U)$ (resp. $\mathcal{X}_{\underline{\underline{A}}}(U)$). It is clear that the family $\{\mathcal{X}^w_{\underline{\underline{A}}}(U);\ U$ open in $A\}$ together with the obvious restrictions $\mathcal{X}^w_{\underline{\underline{A}}}(U) \longrightarrow \mathcal{X}^w_{\underline{\underline{A}}}(V)$ (for $V \subset U$) is a sheaf on A, denoted $\mathcal{X}^w_{\underline{\underline{A}}}$. The sheaf $\mathcal{X}_{\underline{\underline{A}}}$ is defined similarly; in fact it is a subsheaf of $\mathcal{X}^w_{\underline{\underline{A}}}$. Both $\mathcal{X}^w_{\underline{\underline{A}}}$ and $\mathcal{X}_{\underline{\underline{A}}}$ have natural structures of $C^\infty_{\underline{\underline{A}}}$-modules and also of sheaves of Lie algebras. For example if $\xi, \eta \in \mathcal{X}^w_{\underline{\underline{A}}}(U)$, then $[\xi, \eta]$ is determined by

$$[\xi, \eta]|X = [\xi|X, \eta|X], \quad X \in \mathcal{A}|U \quad .$$

Any $\xi \in \mathcal{X}^w_{\underline{\underline{A}}}(U)$ may be viewed as a derivation of $C^\infty_{\underline{\underline{A}}}(U)$ as follows

$$(\xi \cdot f)|X = (\xi|X) \cdot (f|X), \quad X \in \mathcal{A}|U,\ f \in C^\infty_{\underline{\underline{A}}}(U) \quad .$$

The fact that $\xi \cdot f \in C^\infty_{\underline{\underline{A}}}(U)$ follows immediately from the control conditions satisfied by ξ and f. The fact that $f \longmapsto \xi \cdot f$ is a derivation of $C^\infty_{\underline{\underline{A}}}(U)$ is obvious. The converse is also true (i.e., any derivation of $C^\infty_{\underline{\underline{A}}}(U)$ is of the above form) but since we shall not use this assertion, we leave its proof to the reader.

Notation. Let $\underline{\underline{A}}$ and $\underline{\underline{B}}$ be w.a.s.'s, $f : \underline{\underline{A}} \dashrightarrow \underline{\underline{B}}$ be a weak morphism, $U \subset A$ and $V \subset B$ be locally closed subsets such that $\underline{\underline{A}}|U$ and $\underline{\underline{B}}|V$ exist, and let $\xi \in \mathcal{X}^w_{\underline{\underline{A}}}(U)$, $\eta \in \mathcal{X}^w_{\underline{\underline{B}}}(V)$ and $\varphi \in C^\infty_{\underline{\underline{A}}}(U)$. The notation

$$df \cdot \xi = \varphi\eta$$

will mean that for any $a \in U \cap f^{-1}(V)$

$$d(f|X) \cdot \xi(a) = \varphi(a)\eta(f(a)) \quad ,$$

where $X \in \mathcal{A}$ is the stratum containing a. For example

$$d\varphi \cdot \xi = (\xi \cdot \varphi) d/dt$$

where d/dt is the canonical vector field on R.

Given two smooth manifolds X and Y, and given $x \in X$ and $y \in Y$, we shall identify $T(X \times Y)_{(x,y)}$ with $TX_x \times TY_y$ as usual. Then for any w.a.s.'s $\underline{\underline{A}}$ and $\underline{\underline{B}}$ and any open subsets $U \subset A$ and $V \subset B$ there exists a canonical map $\mathcal{X}^w_{\underline{\underline{A}}}(U) \times \mathcal{X}^w_{\underline{\underline{B}}}(V) \longrightarrow \mathcal{X}^w_{\underline{\underline{A}} \times \underline{\underline{B}}}(U \times V)$, $(\xi, \eta) \longmapsto \xi \times \eta$, where $(\xi \times \eta)(x, y) = (\xi(x), \eta(y))$.

2.2. Let $\underline{\underline{A}}$ be a w.a.s. and $\xi \in \mathcal{X}^w_{\underline{\underline{A}}}(A)$. For $X \in \mathcal{A}$ and $a \in X$, let $\lambda_a : (s^\xi_a, t^\xi_a) \longrightarrow X$ be the maximal integral curve of $\xi|X$ through a (i.e., $0 \in (s^\xi_a, t^\xi_a) \subset R$, $\lambda_a(0) = a$, $\xi(\lambda_a(s)) = d\lambda_a \cdot (\frac{d}{dt}|_{t=s})$ for $s \in (s^\xi_a, t^\xi_a)$ and if $\lambda' : (s', t') \longrightarrow X$ has the above two properties, then $(s', t') \subset (s^\xi_a, t^\xi_a)$ and $\lambda' = \lambda_a|(s', t')$; the property of being maximal is equivalent to the following: if $t^\xi_a < \infty$ (resp. $s^\xi_a > -\infty$) and (s_n) is a sequence contained in (s^ξ_a, t^ξ_a) and converging to t^ξ_a (resp. s^ξ_a), then the sequence $(\lambda_a(s_n))$ does not converge in X). Define $D_\xi \subset A \times R$ and $\lambda_\xi : D_\xi \longrightarrow A$ as follows

$$D_\xi = \{(a, t) \in A \times R; \quad t \in (s^\xi_a, t^\xi_a)\}$$

$$\lambda_\xi(a, t) = \lambda_a(t) \quad .$$

λ_ξ is called the flow associated to ξ. For any $t \in R$ denote $D_\xi^t = \{a \in A;\ (a, t) \in D_\xi\}$ and let $\lambda_\xi^t : D_\xi^t \longrightarrow A$ be given by $\lambda_\xi^t(a) = \lambda_\xi(a, t)$. From the properties of the flows of smooth vector fields on smooth manifolds it follows that if $(a, t) \in D_\xi$ and $b = \lambda_\xi(a, t)$ then $s_b^\xi = s_a^\xi - t$, $t_b^\xi = t_a^\xi - t$ and for any $s \in (s_b^\xi, t_b^\xi)$ one has

(2.2.1) $$\lambda_\xi(\lambda_\xi(a, t), s)) = \lambda_\xi(a, t+s) \quad .$$

2.3. LEMMA. Let $\underline{\underline{A}}$ be an a.s. and let $\xi \in \mathcal{X}_{\underline{\underline{A}}}(A)$. Then

(i) let $y \in A$; if $t_y^\xi < \infty$ (resp. $s_y^\xi > -\infty$) and (s_n) is a sequence in (s_y^ξ, t_y^ξ) converging to t_y^ξ (resp. s_y^ξ), the sequence $(\lambda_\xi(y, s_n))$ does not converge in A;

(ii) D_ξ is open in $A \times R$ and λ_ξ is a morphism from $\underline{\underline{A}} \times \underline{\underline{R}}|D_\xi$ to $\underline{\underline{A}}$;

(iii) for any $t \in R$, D_ξ^t is open in A and λ_ξ^t is an isomorphism of $\underline{\underline{A}}|D_\xi^t$ on $A|D_\xi^{-t}$, its inverse being λ_ξ^{-t}.

Proof. Let $X \in \mathcal{A}$ and $x \in X$. Choose a compact neighborhood K of x in X and $t_1, t_2 \in R$ such that $s_x^\xi < t_1 < 0 < t_2 < t_x^\xi$ and $K \times [t_1, t_2] \subset D_{\xi|X}$. Choose also ε such that all the control conditions involved hold on T_X^ε. Set $\delta = \min\{\varepsilon(a);\ a \in \lambda_\xi(K \times [t_1, t_2])\}$ and $V = \{a \in T_X^\varepsilon;\ \pi_X(a) \in K,\ \rho_X(a) < \delta/2\}$. Let $a \in V$. Then, for a sufficiently small $t' > 0$ and any t with $|t| < t'$

(2.3.1) $$\lambda_\xi(a, t) \in T_X^\varepsilon \quad .$$

Since $d\pi_X \cdot (\xi | T_X^\varepsilon) = \xi | X$ and $d\rho_X \cdot (\xi | T_X^\varepsilon) = 0$, we can shrink t' such that

(2.3.2) $$\pi_X(\lambda_\xi(a, t)) = \lambda_\xi(\pi_X(a), t)$$

and

(2.3.3) $$\rho_X(\lambda_\xi(a, t)) = \rho_X(a)$$

for $|t| < t'$. If $t' < \min\{t_a^\xi, t_2\}$, then (2.3.1) is also true for $t = t'$

(use the fact that $(\pi_X, \rho_X) : T_X^\varepsilon \longrightarrow X \times [0\ \varepsilon)$ is proper). It follows that (2.3.2) and (2.3.3) are true for $t = t'$ too. Now, if $s > 0$ is small enough, $\lambda_\xi(a, t'+s) \in T_X^\varepsilon$ and

$$\pi_X(\lambda_\xi(a, t'+s)) = \pi_X(\lambda_\xi(\lambda_\xi(a, t'), s)) = \lambda_\xi(\pi_X(\lambda_\xi(a, t')), s)$$
$$= \lambda_\xi(\lambda_\xi(\pi_X(a), t'), s) = \lambda_\xi(\pi_X(a), t'+s) \quad ;$$

similarly

$$\rho_X(\lambda_\xi(a, t'+s)) = \rho_X(a) \quad .$$

We deduce that (2.3.1), (2.3.2) and (2.3.3) are valid for $0 \leq t < \min\{t_a^\xi, t_2\}$. The same arguments apply for negative t's and therefore

(2.3.4) (2.3.1), (2.3.2) and (2.3.3) hold for $\max\{s_a^\xi, t_2\} < t < \min\{t_a^\xi, t_2\}$.

Let us prove (i). Assume it is not true. To make a choice, let $\lim s_n = t_y^\xi < \infty$ and let $x = \lim y_n$, where $y_n = \lambda_\xi(y, s_n)$. Let X and Y be the strata of $\underline{\underline{A}}$ which contains x and y respectively. Then $y_n \in Y$ for all n and therefore $X \leq Y$. Since (y_n) cannot converge to a point in Y (because of the maximality of λ_y, see 2.2) it follows that $X < Y$. There is no loss of generality in assuming the sequence (s_n) increasing. For large n and any $p > 0$, $0 \leq s_{n+p} - s_n < t_y^\xi - s_n = t_{y_n}^\xi$, $0 \leq s_{n+p} - s_n < t_2$ and $y_n \in V$. By (2.3.4) and (2.3.3)

$$0 < \rho_X(y_n) = \rho_X(\lambda_\xi(y_n, s_{n+p} - s_n)) = \rho_X(\lambda_\xi(\lambda_\xi(y, s_n), s_{n+p} - s_n))$$
$$= \rho_X(\lambda_\xi(y, s_{n+p})) = \rho_X(y_{n+p}) \quad ;$$

thus the sequence $(\rho_X(y_n))$ cannot converge to zero. Since $\rho_X(x) = 0$, this contradicts the continuity of ρ_X. Thus (i) is true.

To prove (ii), let $(x, t) \in D_\xi$, let the notation be as in the first part of the proof and choose t_1 and t_2 with the additional property that $t_1 < t < t_2$. Let $a \in V$ and suppose that $t_a^\xi < t_2$. Then (by (2.3.4)) (2.3.1), (2.3.2) and (2.3.3) are valid for any $0 \le t < t_a^\xi$. Let (s_n) be a sequence in $(0, t_a^\xi)$ converging to t_a^ξ. Set $a_n = \lambda_\xi(a, t_n)$. By (2.3.2)

$$\lim \pi_X(a_n) = \lambda_\xi(\pi_X(a), t_a^\xi)$$

and by (2.3.3)

$$\lim \rho_X(a_n) = \rho_X(a) \quad .$$

Since $(\pi_X, \rho_X) : T_X^\varepsilon \longrightarrow X \times [0, \varepsilon)$ is proper, there exists a subsequence of (a_n) which converges. This contradicts (i). It follows that $t_a^\xi \ge t_2$. Similarly $s_a^\xi \le t_1$. Thus $V \times (t_1, t_2) \subset D_\xi$ which proves that D_ξ is open in $A \times R$. In view of (2.3.4) we deduce that (2.3.2) and (2.3.3) hold for any $(a, t) \in V \times (t_1, t_2)$. This proves that λ_ξ is a morphism (see the last remark in 1.2.4). Finally (iii) is a trivial consequence of (ii). Q.E.D.

2.4. LEMMA. Let $\underline{\underline{A}}$ be an a.s., M be a smooth manifold and $f : A \longrightarrow M$ be a controlled submersion. Let also $A_0 \subset A$ be a closed subset which is a union of strata (thus $\underline{\underline{A}}|A_0$ exists; we shall denote it $\underline{\underline{A}}_0$), ζ be a smooth vector field on M and $\xi_0 \in \mathcal{X}_{\underline{\underline{A}}_0}(A_0)$ be such that $df \cdot \xi_0 = \zeta$. Then there exists $\xi \in \mathcal{X}_{\underline{\underline{A}}}(A)$ such that $df \cdot \xi = \zeta$ and $\xi | A_0 = \xi_0$. If f is proper then $D_\xi = (f \times 1_R)^{-1}(D_\zeta)$.

<u>Proof</u>. We begin with the first assertion. Because of the existence of controlled partitions of unity the problem is local and therefore it suffices to consider only the case when $\underline{\underline{A}}$ is of finite depth. If $\operatorname{depth}(\underline{\underline{A}}) = 0$, then A is a smooth manifold, A_0 is the union of some connected components of A and the assertion is obvious. Assume inductively that the assertion is true whenever the depth of the domain (of the controlled submersion) is less than $\operatorname{depth}(\underline{\underline{A}})$. Let $a \in A$ and $X \in \mathcal{A}$ be the stratum which contains a. Let ξ_X be a smooth vector field on X such that

(2.4.1) $$df \cdot \xi_X = \zeta \quad ;$$

if $X \subset A_0$, we take $\xi_X = \xi_0 | X$. Choose now a sufficiently small ε such that all the control conditions involved are verified on T_X^ε (in particular if $T_X^\varepsilon \cap A_0 \neq \phi$, then $X \subset A_0$; also the control conditions involved in the definition of ξ_0 are verified on $T_X^\varepsilon \cap A_0$). It is clear that $\operatorname{depth}(\underline{\underline{A}} | T_X^\varepsilon \setminus X) < \operatorname{depth}(\underline{\underline{A}})$ and $(\pi_X, \rho_X) | T_X^\varepsilon \setminus X : T_X^\varepsilon \setminus X \longrightarrow X \times R$ is a controlled submersion. Thus, by induction, there exists $\eta \in \mathcal{X}_{\underline{\underline{A}}}(T_X^\varepsilon \setminus X)$ such that

(2.4.2) $$d\pi_X \cdot \eta = \xi_X, \quad d\rho_X \cdot \eta = 0$$

and

$$\eta | A_0 \cap (T_X^\varepsilon \setminus X) = \xi_0 | A_0 \cap (T_X^\varepsilon \setminus X) \quad .$$

But (2.4.2) simply means that $\xi^a = \{\xi^a(b);\ b \in T_X^\varepsilon\}$ given by

$$\xi^a(b) = \begin{cases} \xi_X(b), & b \in X \\ \eta(b), & b \in T_X^\varepsilon \setminus X \end{cases}$$

is a controlled vector field on T_X^ε. Clearly $\xi^a | A_0 \cap T_X^\varepsilon = \xi_0 | A_0 \cap T_X^\varepsilon$; (1.1.6), (2.4.1) and (2.4.2) show now that

(2.4.3) $$df \cdot \xi^a = \zeta \quad .$$

Thus for any $a \in A$ there exist an open neighborhood U_a of a in A and $\xi^a \in \mathcal{X}_{\underline{\underline{A}}}(U_a)$ which extends $\xi_0 | A_0 \cap U_a$ and verifies (2.4.3). A partition of unity argument completes the proof of the first statement.

To prove the second one, let $X \in \mathcal{A}$, $x \in X$ and $y = f(x)$. Since $df \cdot \xi = \zeta$, it follows that $(s_x^\xi, t_x^\xi) \subset (s_y^\zeta, t_y^\zeta)$. Assume that $t_x^\xi < t_y^\zeta$. Then, because f is proper, there exists a sequence (s_n) in $(0, t_x^\xi)$ converging to t_x^ξ and $(\lambda_\xi(x, s_n))$ converging to some $a \in A$. This contradicts Lemma 2.3(i). Thus $t_x^\xi = t_y^\zeta$ and similarly $s_x^\xi = s_y^\zeta$. This proves that $D_\xi = (f \times 1_R)^{-1}(D_\zeta)$. Q.E.D.

As an application of the above lemmas we shall prove a criterion for the "local triviality" of certain controlled submersions. First we need two more definitions.

2.5. Let $\underline{\underline{A}}$ be an a.s. and M be a smooth manifold. A controlled map $f : A \longrightarrow M$ is called trivial if there exist an a.s. $\underline{\underline{A}}_0$ and an isomorphism $F : \underline{\underline{A}}_0 \times \underline{M} \longrightarrow \underline{\underline{A}}$ such that $f(F(a, x)) = x$.

More generally, f is called locally trivial if any $x \in M$ has an open neighborhood U such that $f|f^{-1}(U) : f^{-1}(U) \longrightarrow U$ is trivial, $f^{-1}(U)$ being endowed with the a.s. structure $\underline{\underline{A}}|f^{-1}(U)$.

2.6. THEOREM (Thom first isotopy lemma). Let $\underline{\underline{A}}$ be an a.s., M be a smooth manifold and $f : A \longrightarrow M$ be a proper controlled submersion. Then f is locally trivial.

Proof. The problem is local with respect to M and therefore we may assume that $M = R^n$ and $0 \in f(A)$. Let $\zeta_i = d/dt_i$, $i = 1, \ldots, n$, be the canonical vector fields on R^n. By Lemma 2.4 there exist $\xi_i \in \mathcal{X}_{\underline{\underline{A}}}(A)$ such that $df \cdot \xi_i = \zeta_i$, $i = 1, \ldots, n$. Let $\lambda_i : D_i \longrightarrow A \times R$ be the flow associated to ξ_i; since f is proper, $D_i = A \times R$. Let $A_0 = f^{-1}(0)$ and define $F : A_0 \times R^n \longrightarrow A$ and $G : A \longrightarrow A_0 \times R^n$ by

$$F(a_0, t_1, \ldots, t_n) = \lambda_n^{t_n}(\lambda_{n-1}^{t_{n-1}}(\ldots(\lambda_1^{t_1}(a_0))\ldots))$$

$$G(a) = (\lambda_1^{-s_1}(\lambda_2^{-s_2}(\ldots(\lambda_n^{-s_n}(a))\ldots)), (s_1, \ldots, s_n))$$

where $(s_1, \ldots, s_n) = f(a)$. It is obvious that F and G are continuous, $G \circ F = 1_{A_0 \times R^n}$, $F \circ G = 1_A$ and $f(F(a_0, (t_1, \ldots, t_n)) = (t_1, \ldots, t_n)$.

Let us check that $\underline{\underline{A}}|A_0$ exists. For any $X \in \mathcal{A}$ choose an $\varepsilon(X) = \varepsilon$ such that all the control conditions involved are verified on T_X^ε and set $T_{X \cap A_0} = T_X^\varepsilon \cap A_0$. Let $a \in T_{X \cap A_0}$. Since f is controlled, $f(\pi_X(a)) = f(a) = 0$ and thus $\pi_X(a) \in X \cap A_0$. Therefore $X \cap A_0 \subset T_{X \cap A_0} \subset A_0 \cap \pi_X^{-1}(X \cap A_0)$. It is obvious that (1.2.3.1) and (1.2.3.3) hold (the last one because f is submersive). Since for any $X \in \mathcal{A}$, $X = F((X \cap A_0) \times R^n)$, (1.2.3.2) is also true. Finally (1.2.3.4) follows from (1.2.1.3) after noticing that for any $X \in \mathcal{A}$ and $x \in X \cap A_0$

$$T(X \cap A_0)_x = \ker((df)|TX_x) \quad .$$

Now set $\underline{\underline{A}}_0 = \underline{\underline{A}}|A_0$. It remains to notice that F and G are morphisms, which is obvious because the λ_i's are morphisms. Q.E.D.

2.7. Remark. Let $\underline{\underline{A}}$ be an a.s. and $X \in \mathcal{A}$. Let ε be sufficiently small such that all the control conditions involved hold on T_X^ε. Then $(\pi_X, \rho_X)|T_X^\varepsilon \setminus X : T_X^\varepsilon \setminus X \longrightarrow X \times (0, \varepsilon)$ is a proper controlled submersion and thus it is locally trivial. If $Y \in \mathcal{A}$ and $X < Y$ it follows that $(\pi_X, \rho_X)|Y \cap T_X^\varepsilon : Y \cap T_X^\varepsilon \longrightarrow X \times (0, \varepsilon)$ is also a locally trivial fibration. Therefore $(\pi_X, \rho_X)(Y \cap T_X^\varepsilon)$ is open and closed in $X \times (0, \varepsilon)$. Since $X < Y$, it follows that

(2.7.1) $(\pi_X, \rho_X)|Y \cap T_X^\varepsilon : Y \cap T_X^\varepsilon \longrightarrow X \times (0, \varepsilon)$ is surjective.

2.8. Let $\underline{\underline{A}}$ be a w.a.s. A locally closed subset $A_0 \subset A$ is called saturated if $\underline{\underline{A}}|A_0$ exists and for any $X \in \mathcal{A}$ with $X \cap A_0 \neq \phi$ there exists an ε such that

(2.8.1) $$\pi_X^{-1}(X \cap A_0) \cap T_X^\varepsilon = A_0 \cap T_X^\varepsilon \quad .$$

Suppose that $A_0 \subset A$ is saturated. Let $X < Y$ be strata of $\underline{\underline{A}}$ and $y \in Y \cap A_0 \cap T_X^\varepsilon$, ε being as above. Then

(2.8.2) $$T(Y \cap A_0)_y = (d(\pi_X|Y \cap T_X^\varepsilon))^{-1}(T(X \cap A_0)_{\pi_X(y)}) \quad .$$

2.9. LEMMA. Let $\underline{\underline{B}}$ be an a.s., $\underline{\underline{A}}$ be a w.a.s., $A_0 \subset A$ be saturated and $f : \underline{\underline{B}} \dashrightarrow \underline{\underline{A}}$ be transverse to A_0, i.e., for any $Y \in \mathcal{B}$ and $X \in \mathcal{A}$ with $f(Y) \subset X$, the smooth map $f|Y : Y \longrightarrow X$ is transverse to $X \cap A_0$. Then $B_0 = f^{-1}(A_0)$ is saturated in $\underline{\underline{B}}$; in particular $\underline{\underline{B}}|B_0$ exists.

Proof. For any $Y \in \mathcal{B}$ choose a $\delta(Y) = \delta$ such that all the control conditions involved hold on T_Y^δ and, if $f(Y) \subset X \in \mathcal{A}$, $X \cap A_0 \neq \phi$ and ε

is as in the definition of a saturated subset (see 2.8), then $f(T^{\delta}_{Y}) \subset T^{\varepsilon}_{X}$. Set $T_{Y\cap B_0} = T^{\delta}_{Y} \cap B_0$ and let $b \in T_{Y\cap B_0}$. Since f is a weak morphism and (2.8.1) holds

$$f(\pi_Y(b)) = \pi_X(f(b)) \in A_0$$

and thus $\pi_Y(b) \in Y \cap B_0$. Therefore $Y \cap B_0 \subset T_{Y\cap B_0} \subset B_0 \cap \pi_Y^{-1}(Y\cap B_0)$. It is clear that (1.2.3.1) and (1.2.3.3) are verified (the last one because of the transversality assumption). Let us check (1.2.3.2). Let $Y, Y' \in \mathcal{B}$ be such that $Y \cap B_0 \neq Y' \cap B_0$ and $Y \cap B_0 \subset c\ell_{B_0}(Y' \cap B_0) \neq \phi$. Then necessarily $Y < Y'$. Let $b \in Y \cap B_0$. By (2.7.1) there exists a sequence (b_n) in $T^{\delta}_{Y} \cap Y'$ such that $\pi_Y(b_n) = b$ and $\rho_Y(b_n) = 1/n$ for n large enough. By (1.1.3) (b_n) converges to b. Now

$$\pi_X(f(b_n)) = f(\pi_Y(b)) = f(b) \in A_0 \quad .$$

Using again (2.8.1) we deduce that $f(b_n) \in A_0$ and thus $b_n \in Y' \cap B_0$. Therefore $Y \cap B_0 \subset c\ell_{B_0}(Y' \cap B_0)$. To check (1.2.3.4) one has to use (2.8.2) and notice that for any $X \in \mathcal{A}$ and $Y \in \mathcal{B}$ with $f(Y) \subset X$, and any $b \in Y \cap B_0$

$$T(Y\cap B_0)_b = (d(f|Y))^{-1}(T(X\cap A_0)_{f(b)}) \quad .$$

Thus $\underline{\underline{B}}|B_0$ exists. At the beginning of the proof we have seen that $B_0 \cap T^{\delta}_{Y} = T_{Y\cap B_0} \subset \pi_Y^{-1}(Y\cap B_0)$. Thus $B_0 \cap T^{\delta}_{Y} \subset \pi_Y^{-1}(Y\cap B_0) \cap T^{\delta}_{Y}$. Let $b \in \pi_Y^{-1}(Y\cap B_0) \cap T^{\delta}_{Y}$. Then

$$\pi_X(f(b)) = f(\pi_Y(b)) \in X \cap A_0$$

and thus $f(b) \in \pi_X^{-1}(X\cap A_0) \cap T^{\varepsilon}_{X}$. By (2.8.1), $f(b) \in A_0$. Therefore $b \in B_0 \cap T^{\delta}_{Y}$ which proves that $B_0 \cap T^{\delta}_{Y} = \pi_Y^{-1}(Y\cap B_0) \cap T^{\delta}_{Y}$. It follows that B_0 is saturated. Q.E.D.

2.10. Remark. The fact that $\underline{\underline{B}}$ was an a.s. was used only once in the above proof, namely in order to apply Remark 2.7. Thus Lemma 2.9 is still true if $\underline{\underline{B}}$ is only a w.a.s. such that for any strata $Y < Y'$ of $\underline{\underline{B}}$, $(\pi_Y, \rho_Y)|T_Y^\delta \cap Y' : T_Y^\delta \cap Y' \longrightarrow Y \times (0, \delta)$ is surjective for some δ.

2.11. LEMMA. Let $\underline{\underline{B}}$ and $\underline{\underline{C}}$ be a.s.'s, $\underline{\underline{A}}$ be a w.a.s., $f : \underline{\underline{B}} \dashrightarrow \underline{\underline{A}}$ and $g : \underline{\underline{C}} \dashrightarrow \underline{\underline{A}}$ be transverse, i.e., for any $Y \in \mathcal{B}$, $Z \in \mathcal{C}$ and $X \in \mathcal{A}$ with $f(Y) \cup g(Z) \subset X$ the smooth maps $f|Y : Y \longrightarrow X$ and $g|Z : Z \longrightarrow X$ are transverse. Denote $B \times_A C = \{(b, c) \in B \times C;\ f(b) = g(c)\}$. Then $\underline{\underline{B}} \times \underline{\underline{C}}|B \times_A C$ exists and is an a.s. if $\operatorname{depth}(\underline{\underline{B}}) \cdot \operatorname{depth}(\underline{\underline{C}}) = 0$.

Proof. Let $\Delta_A = \{(a, a') \in A \times A;\ a = a'\}$ and notice that $B \times_A C = (f \times g)^{-1}(\Delta_A)$, where $f \times g : B \times C \longrightarrow A \times A$ is the obvious map. The assertion will follow from Remark 2.10 if we can prove that

(i) Δ_A is saturated in $A \times A$;

(ii) $f \times g$ is transverse to Δ_A.

To prove (i), notice first that $(\mathcal{A} \times \mathcal{A})|\Delta_A = \{\Delta_X;\ X \in \mathcal{A}\}$. For any $X \in \mathcal{A}$ set $T_{\Delta_X} = (T_X \times T_X) \cap \Delta_A$; all the conditions in 1.2.3 are easily checked and thus $\underline{\underline{A}} \times \underline{\underline{A}}|\Delta_A$ exists. Since Δ_A also verifies (2.8.1), it follows that Δ_A is saturated. (ii) is equivalent to the fact that f and g are transverse. Q.E.D.

In the following we shall denote $\underline{\underline{B}} \times \underline{\underline{C}}|B \times_A C$ by $\underline{\underline{B}} \times_A \underline{\underline{C}}$; it is called the fibre product of $\underline{\underline{B}}$ and $\underline{\underline{C}}$ over $\underline{\underline{A}}$ (with respect to f and g).

2.12. Let $\underline{\underline{A}}$ be an a.s., $X \in \mathcal{A}$ be a stratum and ε be such that all the conditions involved hold on $T_X^{\varepsilon'}$ for some $\varepsilon' > \varepsilon$. Then $\underline{\underline{A}}|S_X^\varepsilon$ exists.

In view of Lemma 2.9 the assertion follows from the following remarks:

(i) set $U = T_X^{\varepsilon'} \setminus X$ and $\underline{\underline{U}} = \underline{\underline{A}}|U$; then $f = (\pi_X, \rho_X)|U : U \longrightarrow X \times (0, \varepsilon')$ is controlled and transverse to the submanifold $X \times \{\varepsilon\}$ of $X \times (0, \varepsilon')$;

(ii) $S_X^\varepsilon = f^{-1}(X \times \{\varepsilon\})$;

(iii) apply (1.2.3.5).

Notes. Most of the notions in this chapter are due to Thom [T_1]; we have adopted the presentation of Mather [M_1] (however the proof of Lemma 2.4 is simpler here: we use another type of induction which concentrates many arguments). The notion of saturated subset (section 2.8) is due to Goresky ([G_2], section 4.1; he uses another terminology).

3. ABSTRACT THOM MAPPINGS

3.1.1. Let V, V_1, V_2 and W be real vector spaces and f_1, f_2, g_1 and g_2 be linear maps such that the diagram

$$\begin{array}{ccc} W & \xrightarrow{f_1} & V_1 \\ {\scriptstyle f_2}\downarrow & & \downarrow{\scriptstyle g_1} \\ V_2 & \xrightarrow[g_2]{} & V \end{array}$$

is commutative. This square is called regular if for any $v_1 \in V_1$ and $v_2 \in V_2$ such that $g_1(v_1) = g_2(v_2)$, there exists $w \in W$ such that $f_1(w) = v_1$ and $f_2(w) = v_2$.

3.1.2. Let

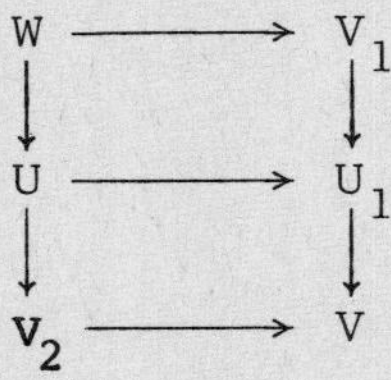

be a commutative diagram of real vector spaces and linear maps. Suppose that the top and bottom squares are regular. It follows immediately that the big square is also regular.

3.1.3. Consider now a commutative square

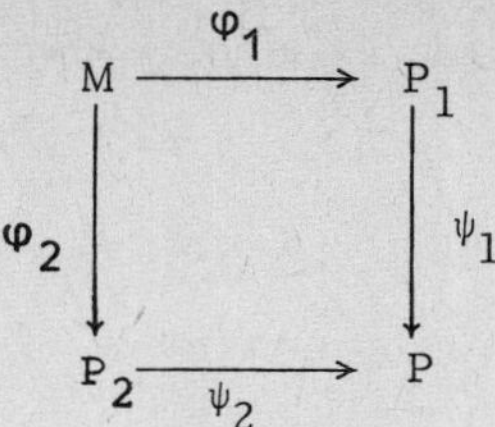

of smooth manifolds and smooth maps. It is called <u>regular</u> if for any $m \in M$

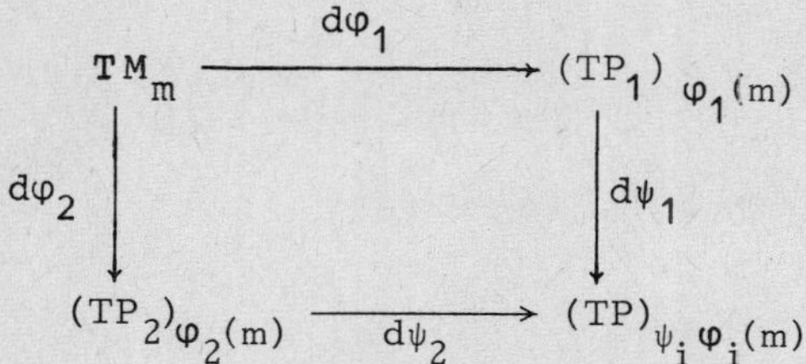

is a regular square of real vector spaces and linear maps.

3.2. Let $\underline{\underline{B}}$ be a w.a.s. and $\underline{\underline{A}}$ be an a.s. A weak morphism $f : \underline{\underline{B}} \dashrightarrow \underline{\underline{A}}$ is called an <u>abstract Thom mapping</u> (a.T.m.), denoted $f : \underline{\underline{B}} \vdash\!\!\longrightarrow \underline{\underline{A}}$, if

(3.2.1) it is submersive;

(3.2.2) for any $X \in \mathcal{A}$ and any $Y < Y' \in \mathcal{B}$ such that $Y, Y' \subset f^{-1}(X)$, the tubes τ_Y and $\tau_{Y'}$ verify (1.1.5);

(3.2.3) for any $Y < Y' \in \mathcal{B}$ with $f(Y) \subset X \in \mathcal{A}$ and $f(Y') \subset X' \in \mathcal{A}$ there exist ε and δ such that $f(T_Y^\delta) \subset T_X^\varepsilon$, $\pi_Y | T_Y^\delta \cap Y'$ and $\pi_X | T_X^\varepsilon \cap X'$ are smooth and the square

$$\begin{array}{ccc} T_Y^\delta \cap Y' & \xrightarrow{\pi_Y} & Y \\ \downarrow f & & \downarrow f \\ T_X^\varepsilon \cap X' & \xrightarrow{\pi_X} & X \end{array}$$

is regular.

Notice that, as a consequence of (3.2.2) and (3.2.1)

(3.2.4) for any $X \in \mathcal{A}$, $\underline{\underline{B}}|f^{-1}(X)$ exists, is an a.s. and $f|f^{-1}(X) : f^{-1}(X) \longrightarrow X$ is a controlled submersion.

Let $f : \underline{\underline{B}} \longmapsto \underline{\underline{A}}$ and $f' : \underline{\underline{B}}' \longmapsto \underline{\underline{A}}'$ be a.T.m.'s, $F : \underline{\underline{A}} \longrightarrow \underline{\underline{A}}'$ be a morphism (resp. an isomorphism) and $G : \underline{\underline{B}} \dashrightarrow \underline{\underline{B}}'$ be a weak morphism (resp. a weak isomorphism). We shall say that the pair (G, F) is a <u>morphism</u> (resp. an <u>isomorphism</u>) <u>from</u> f <u>to</u> f' (or that G is a <u>morphism</u> (resp. an <u>isomorphism</u>) <u>over</u> F) if $f' \circ G = F \circ f$ and for any $X \in \mathcal{A}$ and $X' \in \mathcal{A}'$ with $F(X) \subset X'$, the restriction of G to $f^{-1}(X)$ is a morphism (resp. an isomorphism) from $\underline{\underline{B}}|f^{-1}(X)$ to $\underline{\underline{B}}'|(f')^{-1}(X')$.

3.3. This subsection contains some examples, remarks and constructions to be used later on.

3.3.1. Let $f : \underline{\underline{B}} \dashrightarrow \underline{\underline{A}}$ be a weak morphism and let $\operatorname{depth}(\underline{\underline{A}}) = 0$. Then f is an a.T.m. if and only if $\underline{\underline{B}}$ is an a.s. and f is submersive.

3.3.2. Let $f : \underline{\underline{B}} \longmapsto \underline{\underline{A}}$ and let $U \subset A$ (resp. $V \subset B$) be either a locally closed subset which is a union of strata or an open subset. Suppose that $f(V) \subset U$. Then $f|V : \underline{\underline{B}}|V \dashrightarrow \underline{\underline{A}}|U$ is an a.T.m.

3.3.3. Let $f_0 : \underline{\underline{B}}_0 \longmapsto \underline{\underline{A}}_0$ be an a.T.m. and $\underline{\underline{C}}$ be an a.s. of depth zero. Then $f_0 \times 1_C : \underline{\underline{B}}_0 \times \underline{\underline{C}} \dashrightarrow \underline{\underline{A}}_0 \times \underline{\underline{C}}$ is an a.T.m.

Consider now an a.T.m. $f : \underline{\underline{B}} \longmapsto \underline{\underline{A}}$ and a controlled map $p : A \longrightarrow M$. We shall say that f is <u>trivial</u> <u>over</u> p if there exist an a.T.m. $f_0 : \underline{\underline{B}}_0 \longmapsto \underline{\underline{A}}_0$ and an isomorphism (G, F) from f to $f_0 \times 1_M$ such that $p = p_2 \circ F$, where $p_2 : A \times M \longrightarrow M$ is the canonical projection.

More generally, the a.T.m. $f : \underline{\underline{B}} \longmapsto \underline{\underline{A}}$ is called <u>locally trivial over the controlled map</u> $p : A \longrightarrow M$ if any $m \in M$ has an open neighborhood U such that $f|(p\circ f)^{-1}(U) : \underline{\underline{B}}|(p\circ f)^{-1}(U) \longmapsto \underline{\underline{A}}|p^{-1}(U)$ is trivial over $p|p^{-1}(U) : p^{-1}(U) \longrightarrow U$.

3.3.4. Let $f : \underline{\underline{B}} \longmapsto \underline{\underline{A}}$ be an a.T.m. and let $X \in \mathcal{A}$ and $Y \in \mathcal{B}$ be strata such that $f(Y) \subset X$. Choose ε and δ such that all the control conditions involved hold on T_X^ε and T_Y^δ. Consider the controlled submersion $\pi_X : T_X^\varepsilon \longrightarrow X$ and the smooth submersion $f|Y : Y \longrightarrow X$; they are transverse and thus we can consider the fibre product $\underline{\underline{Y}} \times_X (\underline{\underline{A}}|T_X^\varepsilon)$ (see 2.11). Let also $\underline{\underline{B}}_\varepsilon^\delta = \underline{\underline{B}}|f^{-1}(T_X^\varepsilon) \cap T_Y^\delta$ and define $g : \underline{\underline{B}}_\varepsilon^\delta \longrightarrow \underline{\underline{Y}} \times_X (\underline{\underline{A}}|T_X^\varepsilon)$ by $g(b) = (\pi_Y(b), f(b))$. A direct verification shows that g is an a.T.m.

3.4. Let $f : \underline{\underline{B}} \longmapsto \underline{\underline{A}}$ be an a.T.m. and $V \subset B$ be either a locally closed subset which is a union of strata or an open subset. We shall denote by $\mathcal{X}_{\underline{\underline{B}}}^f(V)$ the subset of $\mathcal{X}_{\underline{\underline{B}}}^w(V)$ consisting of those η which verify

(3.4.1) for any $X \in \mathcal{A}$, $\eta|f^{-1}(X) \cap V$ is a controlled vector field on $\underline{\underline{B}}|f^{-1}(X) \cap V$.

It is obvious that the collection $\{\mathcal{X}_{\underline{\underline{B}}}^f(V);\ V \text{ open in } B\}$ is a subsheaf of the sheaf $\mathcal{X}_{\underline{\underline{B}}}^w$; we shall denote it $\mathcal{X}_{\underline{\underline{B}}}^f$. In fact $\mathcal{X}_{\underline{\underline{B}}}^f$ is a $C_{\underline{\underline{B}}}^\infty$-submodule of $\mathcal{X}_{\underline{\underline{B}}}^w$ and also a subsheaf of Lie algebras of $\mathcal{X}_{\underline{\underline{B}}}^w$.

3.5. LEMMA. Let $f : \underline{\underline{B}} \longmapsto \underline{\underline{A}}$ be an a.T.m., $\xi \in \mathcal{X}_{\underline{\underline{A}}}^w(A)$, $B_0 \subset B$ be a closed union of strata and $\eta_0 \in \mathcal{X}_{\underline{\underline{B}}}^f(B_0)$. Suppose that $df \cdot \eta_0 = \xi$. Then there exists $\eta \in \mathcal{X}_{\underline{\underline{B}}}^f(B)$ such that $\eta|B_0 = \eta_0$ and $df \cdot \eta = \xi$.

Proof. Because of the existence of controlled partitions of unity, the assertion is local. Thus we may assume that $\operatorname{depth}(\underline{\underline{B}})$ is finite. If $\operatorname{depth}(\underline{\underline{B}}) = 0$, then there are no incidence relations between the strata of $\underline{\underline{B}}$ and the assertion is trivial. Assume that $\operatorname{depth}(\underline{\underline{B}}) > 0$ and that the assertion is true for any a.T.m. whose domain has depth less than $\operatorname{depth}(\underline{\underline{B}})$. Let $b \in B$, $Y \in \mathcal{B}$ be the stratum which contains b and $X \in \mathcal{A}$ be the stratum which contains $f(Y)$. Let $g : \underline{\underline{B}}_\varepsilon^\delta \longmapsto \underline{\underline{Y}} \times_X (\underline{\underline{A}}|T_X^\varepsilon)$ be the a.T.m. constructed in 3.3.4. Let also $\underline{\underline{B}}_X = \underline{\underline{B}}|f^{-1}(X)$ and $f_X = f|B_X : B_X \longrightarrow X$. By (3.2.4) f_X is a controlled submersion. As a consequence of Lemma 2.4, there exists

$\zeta \in \mathcal{X}_{\underline{\underline{B}}_X}(B_X)$ such that

$$\text{(3.5.1)} \qquad df \cdot \zeta = \xi$$

and

$$\text{(3.5.2)} \qquad \zeta | B_X \cap B_0 = \eta_0 | B_X \cap B_0 \quad .$$

Let $B_1 = B_0 \cup B_X$; it is a closed union of strata and, as a consequence of (3.5.2), ζ and η_0 determine a vector field $\eta_1 \in \mathcal{X}^f_{\underline{\underline{B}}}(B_1)$ such that $df \cdot \eta_1 = \xi$.

Let $\underline{\underline{C}} = \underline{\underline{B}}^\delta_\varepsilon | (B^\delta_\varepsilon \setminus Y)$, $C_0 = C \cap B_1$, $g' = g|C : \underline{\underline{C}} \longmapsto \underline{\underline{Y}} \times_X (\underline{\underline{A}}|T^\varepsilon_X)$, $\tilde{\eta}_0 = \eta_1 | C_0$ and $\tilde{\xi} = (\zeta|Y) \times (\xi | T^\varepsilon_X) \in \mathcal{X}_{\underline{\underline{Y}} \times \underline{\underline{A}}}(Y \times T^\varepsilon_X)$. Notice that actually $\tilde{\xi}$ is a controlled vector field on $\underline{\underline{Y}} \times_X (\underline{\underline{A}}|T^\varepsilon_X)$ and $dg' \cdot \tilde{\eta}_0 = \tilde{\xi}$. Since $\operatorname{depth}(\underline{\underline{C}}) < \operatorname{depth}(\underline{\underline{B}})$, we may apply the inductive hypothesis to g', $\tilde{\xi}$ and $\tilde{\eta}_0$. Thus there exists $\tilde{\eta} \in \mathcal{X}^{g'}_{\underline{\underline{C}}}(C) \subset \mathcal{X}^f_{\underline{\underline{B}}}(C)$ such that $\tilde{\eta}|C_0 = \tilde{\eta}_0$ and $dg' \cdot \tilde{\eta} = \tilde{\xi}$. The last equality is equivalent to $df \cdot \tilde{\eta} = \xi$ and $d\pi_Y \cdot \tilde{\eta} = \zeta | Y$. Now it is obvious that $\tilde{\eta}$ and $\zeta|Y$ determine an $\eta^b \in \mathcal{X}^f_{\underline{\underline{B}}}(B^\delta_\varepsilon)$ such that $df \cdot \eta^b = \xi$ and $\eta^b | B^\delta_\varepsilon \cap B_0 = \eta_0 | B^\delta_\varepsilon \cap B_0$.

Thus for any $b \in B$ there exist an open neighborhood U^b of b in B and $\eta^b \in \mathcal{X}^f_{\underline{\underline{B}}}(U^b)$ such that $df \cdot \eta^b = \xi$ and $\eta^b | U^b \cap B_0 = \eta_0 | U^b \cap B_0$. A partition of unity argument completes the proof of the lemma. Q.E.D.

3.6. LEMMA. Let $f : \underline{\underline{B}} \longmapsto \underline{\underline{A}}$ be an a.T.M., $\xi \in \mathcal{X}_{\underline{\underline{A}}}(A)$ and $\eta \in \mathcal{X}^f_{\underline{\underline{B}}}(B)$. Suppose that $df \cdot \eta = \xi$. Then

(i) let $y \in B$; if a sequence (s_n) in (s^η_y, t^η_y) converges to $t^\eta_y < \infty$ (resp. $s^\eta_y > -\infty$), the sequence $(\lambda_\eta(y, s_n))$ does not converge in B;

(ii) D_η is open in $B \times R$ and $(\lambda_\eta, \lambda_\xi)$ is a morphism from $(f \times 1_R)|D_\eta : (\underline{\underline{B}} \times \underline{\underline{R}})|D_\eta \longmapsto (\underline{\underline{A}} \times \underline{\underline{R}})|D_\xi$ to f;

(iii) for any $t \in R$, $(\lambda^t_\eta, \lambda^t_\xi)$ is an isomorphism from $f|D^t_\eta : \underline{\underline{B}}|D^t_\eta \longmapsto \underline{\underline{A}}|D_\xi$ to $f|D^{-t}_\eta : \underline{\underline{B}}|D^{-t}_\eta \longmapsto \underline{\underline{A}}|D^{-t}_\xi$;

(iv) if f is proper, $D_\eta = (f \times 1_R)^{-1}(D_\xi)$.

Proof. (i) Assume the contrary. To make a choice, let $\lim s_n = t^\eta_y < \infty$ and $\lim y_n = y'$, where $y_n = \lambda_\eta(y, s_n)$. Let $x = f(y)$, $x' = f(y')$ and let X, X', Y and Y' be the strata containing x, x', y and y' respectively. Clearly $X' \leq X$ and $Y' < Y$. If $X = X'$, then Y and Y' are contained in $f^{-1}(X)$. Using (3.2.4), (3.4.1) and Lemma 2.3(i) we obtain a contradiction. Thus $X' < X$. Since $df \cdot \eta = \xi$, it follows that $(s^\eta_y, t^\eta_y) \subset (s^\xi_x, t^\xi_x)$ and $f(\lambda_\eta(y, s_n)) = \lambda_\xi(x, s_n)$ for any n. If $t^\eta_y < t^\xi_x$, then $x' = \lim \lambda_\xi(x, s_n) = \lambda_\xi(x, t^\eta_y) \in X$; since $X < X'$, this is not possible. Therefore $t^\eta_y = t^\xi_x$. In this case (s_n) converges to $t^\xi_x < \infty$ and $(\lambda_\xi(x, s_n))$ converges to $x' \in A$, which contradicts Lemma 2.3(i). This last contradiction shows that (i) is true.

(ii) Let $(y, t) \in D_\eta$, W_0 be a neighborhood of $\lambda_\eta(y, t)$ in B, $Y \in \mathcal{B}$ be the stratum which contains y, $x = f(y)$ and $X \in \mathcal{A}$ be the stratum which contains x. Since $df \cdot \eta = \xi$, $(x, t) \in D_\xi$. To make a choice, assume that $t \geq 0$. Choose now a compact neighborhood K of y in Y, real numbers t_1, t_2, t'_1, ε_0 and δ, and smooth maps $\varepsilon : X \longrightarrow R^*_+$ and $\delta(Z) : Z \longrightarrow R^*_+$ $(Z \in \mathcal{B})$ such that $t_1 < 0$, $t_2, \varepsilon, \delta > 0$, $s^\eta_y < t_1 \leq t'_1 < t < t_2 < t^\eta_y$ and

(3.6.1) $K \times [t_1, t_2] \subset D_\eta$ (hence $f(K) \times [t_1, t_2] \subset D_\xi$) ;

(3.6.2) all the control conditions involved hold on T^ε_X and $T^{\delta(Z)}_Z$, $Z \in \mathcal{B}$;

(3.6.3) $\delta < \min\{\delta(Y)(b);\ b \in \lambda_\eta(K \times [t_1, t_2])\}$ and
$\varepsilon_0 < \min\{\varepsilon(a);\ a \in \lambda_\xi(f(K) \times [t_1, t_2])\}$;

(3.6.4) if $V_0 = \{b \in T_Y;\ \pi_Y(b) \in K,\ \rho_Y(b) \leq \delta\}$, then $f(V_0) \times [t_1, t_2] \subset D_\xi$;

(3.6.5) if $U = \{a \in T_X;\ \pi_X(a) \in \lambda_\xi(f(K) \times [t_1, t_2]),\ \rho_X(a) \leq \varepsilon_0\}$,
$V = \{b \in T_Y;\ \pi_Y(b) \in \lambda_\eta(K \times [t_1, t_2]),\ \rho_Y(b) \leq \delta\}$ and
$\mathcal{B}^+_Y = \{Z \in \mathcal{B};\ f(Z) \subset X,\ Z > Y\}$, then

$$f^{-1}(U) \cap V \subset T^{\delta/2}_Y \cup \Big(\bigcup_{Z \in \mathcal{B}^+_Y} T^{\delta(Z)}_Z \Big) \quad ;$$

(3.6.6) $\{b \in T_Y^\delta ;\ \pi_Y(b) \in \lambda_\eta(K \times [t_1', t_2])\} \subset W_0'$.

There are no problems with (3.6.1), (3.6.2) and (3.6.3). Since D_ξ is open in $A \times R$ and K is compact, (1.1.3) implies that (3.6.4) is true for δ sufficiently small. Since $D_{\eta|Y}$ is open in $Y \times R$ and $\lambda_{\eta|Y}$ is continuous on $D_{\eta|Y}$, (1.1.3) implies that (3.6.6) is true for δ and $t_2 - t_1'$ sufficiently small. Finally, since V is compact, similar arguments imply that (3.6.5) is true for a sufficiently small ε_0. Notice that, by (3.6.3),

(3.6.7) $$U \subset T_X^\varepsilon \quad \text{and} \quad V \subset T_Y^{\delta(Y)}$$

and, by (1.1.2), U and V are compact. If $(b, s) \in D_\eta$ and $\lambda_\eta(b, s) \in T_Y^{\delta(Y)}$ then

$$\pi_Y(\lambda_\eta(b, s)) = \lambda_\eta(\pi_Y(b), s)$$

and

$$f(\lambda_\eta(b, s)) = \lambda_\xi(f(b), s) \quad .$$

In the following we shall use these relations without any other mention.

Let $b \in T_Y^{\delta(Y)} \cap (\bigcup_{Z \in \mathcal{B}_Y^+} T_Z^{\delta(Z)})$. Then, for $|s|$ sufficiently small,

(*) $$\rho_Y(\lambda_\eta(b, s)) = \rho_Y(b) \quad .$$

Indeed, if $b \in T_Z^{\delta(Z)}$ with $Z \in \mathcal{B}_Y^+$, then

$$\rho_Y(\lambda_\eta(b, s)) = \rho_Y(\pi_Z(\lambda_\eta(b, s))) = \rho_Y(\lambda_\eta(\pi_Z(b), s))$$
$$= \rho_Y(\pi_Z(b)) = \rho_Y(b)$$

(the first and fourth equality are consequences of (3.2.2), the third one is a consequence of (3.4.1) and the second one follows from (2.1.2)).

Assume now that $\lambda_\eta(b, \tilde{s}) \in T_Y^{\delta(Y)} \cap (\bigcup_{Z \in \mathcal{B}_Y^+} T_Z^{\delta(Z)})$. Then from (*) and (2.2.1) we obtain

(3.6.8) $\rho_Y(\lambda_\eta(b, s)) = \rho_Y(\lambda_\eta(b, \tilde{s}))$, s near $\tilde{s}$.

Let $W = \mathrm{int}_B(V_0 \cap f^{-1}(U))$. Clearly $y \in W$. To prove that D_η is open in $B \times R$ it suffices to check that $W \times (t_1, t_2) \subset D_\eta$. Let $b \in W$ and $Z_0 \in \mathcal{B}$ be the stratum which contains b. Suppose that $t_b^\eta < t_2$. Since $\lambda_{\eta|Z_0}$ is continuous, there exists $0 < s_0 < t_b^\eta$ such that

(3.6.9) $$\lambda_\eta(b, s) \in V \cap f^{-1}(U)$$

for $0 \leq s < s_0$ and s_0 is maximal with this property. If $s_0 < t_b^\eta$, since $V \cap f^{-1}(U)$ is closed in B,

$$\lambda_\eta(b, s_0) \in V \cap f^{-1}(U) \subset T_Y^{\delta/2} \cup \left(\bigcup_{Z \in \mathcal{B}_Y^+} T_Z^{\delta(Z)} \right) .$$

If $\lambda_\eta(b, s_0) \in T_Y^{\delta/2}$, since $s_0 < t_b^\eta < t_2$, using the continuity of $\lambda_{\eta|Z_0}$, (3.6.7), (3.6.4) and (3.6.2) we deduce that (3.6.9) is true for $0 \leq s < s'$ with $s' > s_0$. If $\lambda_\eta(b, s_0) \notin T_Y^{\delta/2}$ then $\lambda_\eta(b, s_0) \in T_Y^{\delta(Y)} \cap (\bigcup_{Z \in \mathcal{B}_Y^+} T_Z^{\delta(Z)})$ and, using (3.6.8), (3.6.2) and (3.6.4) we reach again the same conclusion as above. Since this contradicts the maximality of s_0, it follows that $s_0 = t_b^\eta$.

Since $V \cap f^{-1}(U)$ is compact, there exists a sequence (s_n) in (s_b^η, t_b^η) converging to $t_b^\eta < \infty$ and such that the sequence $(\lambda_\eta(b, s_n))$ converges in B. This contradicts (i). Therefore $t_b^\eta \geq t_2$. Similarly $s_b^\eta \leq t_1$. Hence $W \times (t_1, t_2) \subset D_\eta$ which, as already mentioned, proves that D_η is open in $B \times R$.

Notice now that (3.6.9) is true for any $t_1 < s < t_2$. From (3.6.2), (3.6.7) and (3.6.6) it follows immediately that $\lambda_\eta(W \times (t_1', t_2)) \subset W_0'$, which proves the continuity of λ_η. The remaining part of (ii) raises no supplementary difficulties and is left to the reader.

(iii) is obvious. (iv) can be proved exactly as the last assertion of Lemma 2.4. Q.E.D.

3.7. THEOREM (Thom second isotopy lemma). Let $f : \underline{\underline{B}} \longmapsto \underline{\underline{A}}$ be a proper a.T.m., M be a smooth manifold and $p : A \longrightarrow M$ be a proper controlled submersion. Then f is locally trivial over p.

Proof. The assertion being local with respect to M, we may assume that $M = R^n$ and $0 \in p(A)$. Then the vector fields d/dt_i on R^n can be lifted to vector fields $\xi_i \in X_{\underline{\underline{A}}}(A)$ which in turn can be lifted to vector fields $\eta_i \in X^f_{\underline{\underline{B}}}(B)$ (thus $df \cdot \eta_i = \xi_i$ and $dp \cdot \xi_i = d/dt_i$). Let $A_0 = p^{-1}(0)$, $B_0 = f^{-1}(A_0)$ and $f_0 = f|B_0 : B_0 \longrightarrow A_0$. As in the proof of Theorem 2.6 the flows of the ξ_i's and η_i's induce homeomorphisms $F : A_0 \times R^n \longrightarrow A$ and $G : B_0 \times R^n \longrightarrow B$ such that

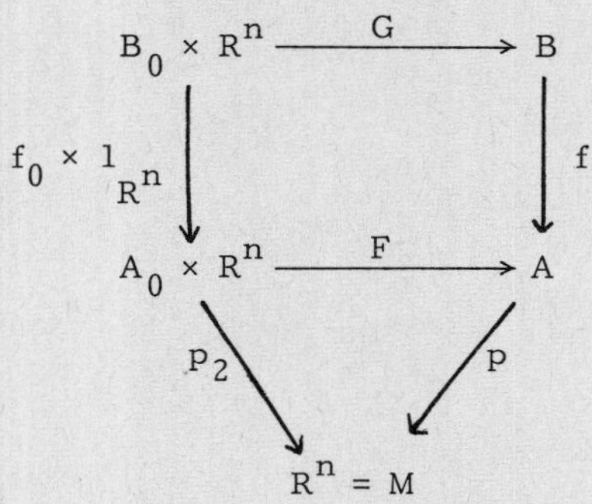

commutes (p_2 is the projection on R^n). We know that $\underline{\underline{A}}|A_0$ exists (see the proof of Theorem 2.6); the same type of arguments show that $\underline{\underline{B}}|B_0$ exists and $f_0 : \underline{\underline{B}}|B_0 \longmapsto \underline{\underline{A}}|A_0$ is an a.T.m. Set $\underline{\underline{B}}_0 = \underline{\underline{B}}|B_0$ and $\underline{\underline{A}}_0 = \underline{\underline{A}}|A_0$. Since (G, F) is obviously an isomorphism from $f_0 \times 1_{R^n} : \underline{\underline{B}}_0 \times \underline{\underline{R}}^n \longmapsto \underline{\underline{A}}_0 \times \underline{\underline{R}}^n$ to $f : \underline{\underline{B}} \longmapsto \underline{\underline{A}}$, the proof is complete. Q.E.D.

Notes. The notions introduced in this chapter are again due to Thom [T_1], while the presentation follows closely that of Mather [M_1]. However the proof of Lemma 3.5 is much simpler here.

4. MANIFOLDS WITH FACES

4.1. Let M be a smooth manifold with corners (see [C]), bM be the boundary of M and $\mathring{M} = M \setminus bM$ be the interior of M (thus it is a smooth manifold without boundary). A <u>face</u> of M consists of a closed subset $B \subset bM$ and a map $F_B : U_B \longrightarrow B \times R_+$, called the <u>collar</u> of the face, such that

(4.1.1) B is a manifold with corners;

(4.1.2) U_B is an open neighborhood of B in M, $F_B(U_B)$ is open in $B \times R_+$ and F_B is a diffeomorphism of U_B onto $F_B(U_B)$;

(4.1.3) $F_B(x) = (x, 0)$ for any $x \in B$.

Notice that

$$F_B(U_B \cap \mathring{M}) = F_B(U_B) \cap (\mathring{M} \times R_+^*) \quad .$$

Define a smooth vector field η_B on U_B by

$$(dF_B)(\eta_B) = (0 \times d/dt)|F_B(U_B) \quad .$$

It is clear that F_B is completely determined by η_B. Let also $p_B : U_B \longrightarrow B$ and $r_B : U_B \longrightarrow R_+$ be given by

$$F_B(x) = (p_B(x), r_B(x)), \; x \in U_B \quad .$$

Then p_B and r_B determine F_B and η_B is the unique vector field on U_B which verifies

(4.1.4) $$dp_B \cdot \eta_B = 0 \quad \text{and} \quad dr_B \cdot \eta_B = d/dt \quad .$$

As in the case of tubes we shall not distinguish between two faces

(B, F_B) and $(B', F_{B'})$ if $B = B'$ and $F_B = F_{B'}$ near B. From now on, in order to simplify the notation, a face (B, F_B) will be denoted simply B; $F_B : U_B \longrightarrow B \times R_+$, η_B, p_B and r_B will always have the above meaning. For formal reasons we shall consider that the empty set is a face of M, with the empty map as collar.

Let now M_1 and M_2 be faces of M. Set $U_i = U_{M_i}$, $F_i = F_{M_i}$, $\eta_i = \eta_{M_i}$, $p_i = p_{M_i}$ and $r_i = r_{M_i}$, $i = 1, 2$. M_1 and M_2 are called <u>compatible</u> if either (a) $M_1 = M_2$ (and $F_1 = F_2$ near M_1), or (b) $M_1 = \phi$, or (c) $M_2 = \phi$, or (d) M_1 and M_2 are non empty, different, $M_1 \cap M_2$ is a face of M_1 (resp. M_2) with collar a suitable restriction of F_2 (resp. F_1) and

(4.1.5) $\quad [\eta_1 | U_1 \cap U_2, \eta_2 | U_1 \cap U_2] = 0$ near $M_1 \cap M_2$;

(4.1.6) $\quad dr_i \cdot \eta_j = 0$ near $M_1 \cap M_2$, $i \neq j$.

As a consequence one also has

(4.1.7) $\quad dp_i \cdot \eta_j = \eta_j | M_i$ near $M_1 \cap M_2$, $i \neq j$;

(4.1.8) $\quad p_1(p_2(x)) = p_2(p_1(x)) \in M_1 \cap M_2$ near $M_1 \cap M_2$;

(4.1.9) $\quad$ there exists an open neighborhood V of $M_1 \cap M_2$ in M such that the assignment $x \longmapsto (p_1(p_2(x)), r_1(x), r_2(x))$ gives rise to a diffeomorphism of V onto an open subset of $(M_1 \cap M_2) \times R_+ \times R_+$.

The smooth manifold with corners M is called a <u>manifold with faces</u> if $bM = \bigcup_{i \in I_M} M_i$, any M_i is a face of M and any two of them are compatible (in view of our last convention it may happen that $M_i = \phi$ for some $i \in I_M$).

<u>Remarks</u>: (1) There exist manifolds with corners which carry no

structure of manifold with faces. An example is given by the following compact domain in R^2:

(2) Let M be a manifold with faces, $I = \{i_1, \dots, i_n\}$ be a finite subset of I_M and $M_I = \bigcap_{i \in I} M_i$. Then M_I is a manifold with faces and there exists an open neighborhood V of M_I in M such that the assignment $x \longmapsto (p_{M_{i_1}}(p_{M_{i_2}}(\dots(p_{M_{i_n}}(x))\dots)), r_{M_{i_1}}(x), \dots, r_{M_{i_n}}(x))$ gives rise to a diffeomorphism of V onto an open subset of $M_I \times R_+ \times \dots \times R_+$.

(3) The family $\{M_i;\ i \in I_M\}$ is locally finite (in M).

4.2. Let M and N be manifolds with faces, $bM = \bigcup_{i \in I_M} M_i$, $bN = \bigcup_{j \in I_N} N_j$. A map $f : M \longrightarrow N$ is called compatible with the faces if $I_M = I^{v,f} \cup I^{h,f}$, $I^{v,f} \subset I_N$ and

(4.2.1) if $i \in I^{v,f}$ then $f^{-1}(N_i) = M_i$ and

$$F_{N_i} \circ f = ((f|M_i) \times 1_{R_+}) \circ F_{M_i} \quad \text{near } M_i \ ;$$

(4.2.2) if $i \in I^{h,f}$ then $f = (f|M_i) \circ p_{M_i}$ near M_i ;

(4.2.3) if $j \in I_N \smallsetminus I^{v,f}$ then $f^{-1}(N_j) = \phi$ or $f^{-1}(N_j) = M$.

A face M_i of M with $i \in I^{v,f}$ (resp. $i \in I^{h,f}$) is called f-vertical (resp. f-horizontal). Given $i \in I^{h,f}$ we shall denote $N_i = N$. For any $i \in I_M$ we shall denote by $f_i : M_i \longrightarrow N_i$ the restriction of f.

Recall that a smooth $f : M \longrightarrow N$, M and N being smooth manifolds with corners, is an embedding if it is locally proper, injective and for any $x \in M$ the differential $df_x : TM_x \longrightarrow TN_{f(x)}$ is injective.

Let now N be a manifold with faces. A subset M of N is called a submanifold of N if there exists a structure of manifold with faces on M such that the inclusion $M \subset N$ is an embedding and is compatible with the faces. In this case M is locally closed in N and the structure of manifold with faces on M, having the above property, is unique.

4.3. Examples and remarks.

4.3.1. Any open subset of a manifold with faces is a submanifold.

4.3.2. Any face of a manifold with faces is a submanifold.

4.3.3. Let M and N be manifolds with faces. Then $M \times N$ has an obvious structure of manifold with faces: its faces are of the form

(a) $M \times N_j$ ($j \in I_N$) with collar $(x, y) \longmapsto (x, p_{N_j}(y), r_{N_j}(y))$: $M \times U_{N_j} \longrightarrow M \times N_j \times R_+$, or

(b) $M_i \times N$ ($i \in I_M$) with collar $(x, y) \longmapsto (p_{M_i}(x), y, r_{M_i}(x))$: $U_{M_i} \times N \longrightarrow M_i \times N \times R_+$.

4.3.4. Let M and N be manifolds with faces, M_0 be a face of M, N_0 be a face of N and $f : M_0 \longrightarrow N_0$ be a diffeomorphism compatible with the faces. Let $M \cup_f N$ be the space obtained from the disjoint sum of M and N by identifying $x \in M_0$ with $f(x) \in N_0$. We shall consider M and N as subsets of $M \cup_f N$. The usual construction can be carried out and we can endow $M \cup_f N$ with a canonical structure of a manifold with faces. Let us describe its faces. If $M_i \neq M_0$ is a face of M such that $M_i \cap M_0 \neq \phi$, then there exists a unique face N_i of N such that

$f(M_i \cap M_0) = N_i \cap N_0$ and $f_i = f|M_i \cap M_0 : M_i \cap M_0 \longrightarrow N_i \cap N_0$ is a diffeomorphism compatible with the faces. We can assume that $f(U_{M_i} \cap M_0) = U_{N_i} \cap N_0$ and that the equality in (4.2.1) holds on $U_{M_i} \cap M_0$ (all the faces of M_0 are f-vertical!). As above we can construct $M_i \cup_{f_i} N_i$; let $U_i = U_{M_i} \cup U_{N_i} \subset M \cup_f N$ and define $F_i : U_i \longrightarrow (M_i \cup_{f_i} N_i) \times R_+$ by

$$F_i(x) = F_{M_i}(x) \in M_i \times R_+ \subset (M_i \cup_{f_i} N_i) \times R_+, \quad x \in U_{M_i},$$

$$F_i(y) = F_{N_i}(y) \in N_i \times R_+ \subset (M_i \cup_{f_i} N_i) \times R_+, \quad y \in U_{N_i}.$$

It is easily seen that F_i is well defined and that $M_i \cup_{f_i} N_i$ is a face of $M \cup_f N$ with collar F_i. The other faces of $M \cup_f N$ are those faces of M (resp. N) which do not intersect M_0 (resp. N_0), their collars in $M \cup_f N$ being just their collars in M (resp. N).

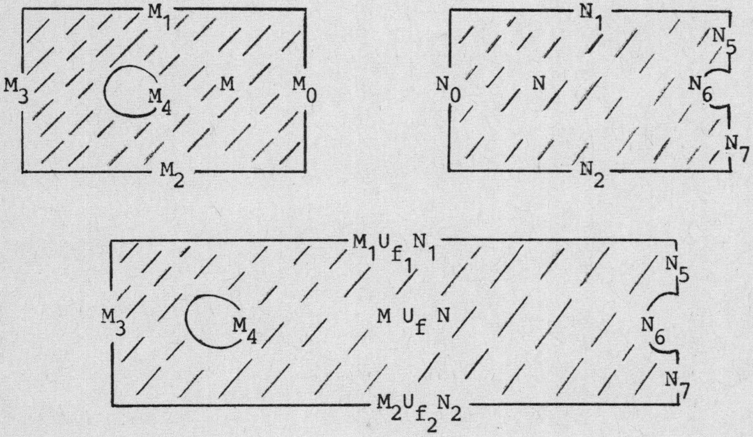

4.3.5. Let M and N be manifolds with faces, $P \subset N$ be a submanifold and $f : M \longrightarrow N$ be a smooth map compatible with the faces. Suppose that f is <u>transverse</u> to P (i.e., for any $x \in M$ with $f(x) \in P$, $TP_{f(x)} + df \cdot (TM_x) = TN_{f(x)}$). Then $f^{-1}(P)$ is a submanifold of M.

We can assume that, for any $j \in I_N$, $P \not\subset N_j$ (otherwise we can replace

M by $f^{-1}(N_j)$, N by N_j and f by $f|f^{-1}(N_j))$. We can also assume that, for any $j \in I_N \setminus I^{v,f}$, $f^{-1}(N_j) = \phi$.

For $i \in I_M$ let $f_i : M_i \longrightarrow N_i$ be the restriction of f (recall that, for $i \in I^{h,f}$, $N_i = N$). Since f is compatible with the faces, f_i is transverse to $P_i = P \cap N_i$. By induction on $\dim(M)$, we may assume that $f_i^{-1}(P_i)$ is a submanifold of M_i. Let $\mathring{f} = f|\mathring{M} : \mathring{M} \longrightarrow \mathring{N}$. Then $\mathring{f}$ is transverse to $\mathring{P} = P \cap \mathring{N}$ and therefore $\mathring{f}^{-1}(\mathring{P})$ is a smooth submanifold of $\mathring{M}$. Now a direct verification shows that there exists a unique structure of a manifold with faces on $f^{-1}(P)$ such that

(1) $f_i^{-1}(P_i)$ is a face of $f^{-1}(P)$, $i \in I_M$;

(2) $\widehat{f^{-1}(P)}^{\circ} = (\mathring{f})^{-1}(P)$

(3) the inclusion $f^{-1}(P) \subset M$ is an embedding compatible with the faces.

4.3.6. The previous construction can be generalized as follows. Let M, N and P be manifolds with faces and $f : M \longrightarrow P$ and $g : N \longrightarrow P$ be smooth maps compatible with the faces. Assume that f and g are <u>transverse</u> (i.e., for any $x \in M$ and $y \in N$ such that $f(x) = g(y)$, $df\cdot(TM_x) + dg\cdot(TN_y) = TP_{f(x)}$). Then $M \times_P N = \{(x, y) \in M \times N;\ f(x) = g(y)\}$ has a canonical structure of a manifold with faces.

There is no loss of generality in assuming that, for any $k \in I_P$, $f(M) \subset P_k$ and $g(N) \not\subset P_k$ (otherwise we can replace P, M and N with P_k, $f^{-1}(P_k)$ and $g^{-1}(P_k)$ respectively).

Let now M_i be an f-vertical face of M, P_i be the corresponding face of P ($M_i = f^{-1}(P_i)$) and $N_i = g^{-1}(P_i)$. Then N_i is either a non empty g-vertical face of N or $N_i = \phi$. Consider the case $N_i \neq \phi$. As above we can define $M_i \times_{P_i} N_i$ (with respect to f_i and g_i, the restrictions of f and g). By induction on $\dim(M)$ we can assume that $M_i \times_{P_i} N_i$ is a manifold with faces. Let $F_i : U_i \longrightarrow M_i \times R_+$ and $G_i : V_i \longrightarrow N_i \times R_+$ be the collars of M_i and N_i respectively; we can assume that the equality in (4.2.1) holds on U_i (resp. V_i). Then $W_i = \{(x, y) \in M \times_P N;$

$x \in U_i$, $y \in V_i\}$ is an open neighborhood of $M_i \times_{P_i} N_i$ in $M \times_P N$ and we can define $H_i : W_i \longrightarrow (M_i \times_P N_i) \times R_+$ by

$$H_i(x, y) = ((p_{M_i}(x), p_{N_i}(y)), r_{M_i}(x))$$

(by (4.2.1), $r_{M_i}(x) = r_{P_i}(f(x)) = r_{N_i}(y)$). It is obvious that $H_i(W_i)$ is open in $(M_i \times_{P_i} N_i) \times R_+$ and that H_i is a homeomorphism of W_i on $H_i(W_i)$.

Next let M_i be an f-horizontal face of M. Then $f|M_i : M_i \longrightarrow P$ and $g : N \longrightarrow P$ are transverse and, again by induction on $\dim(M)$, we can assume that $M_i \times_P N$ is a manifold with faces. Let $F_i : U_i \longrightarrow M_i \times R_+$ be as above and set $\tilde{U}_i = \{(x, y) \in M \times_P N;\ x \in U_i\}$. Define

$\tilde{F}_i : \tilde{U}_i \longrightarrow (M_i \times_P N) \times R_+$ by

$$\tilde{F}_i(x, y) = ((p_{M_i}(x), y), r_{M_i}(x)) \quad .$$

Then $\tilde{U}_i$ is an open neighborhood of $M_i \times_P N$ in $M \times_P N$, $\tilde{F}_i(\tilde{U}_i)$ is open in $(M_i \times_P N) \times R_+$ and $\tilde{F}_i$ is a homeomorphism of $\tilde{U}_i$ on $\tilde{F}_i(\tilde{U}_i)$.

The same construction applies to a g-horizontal face N_j of N and we can define $\tilde{G}_j : \tilde{V}_j \longrightarrow (M \times_P N_j) \times R_+$ with similar properties.

Finally $f|\mathring{M} : \mathring{M} \longrightarrow \mathring{P}$ and $g|\mathring{N} : \mathring{N} \longrightarrow \mathring{P}$ are transverse and thus $\mathring{M} \times_{\mathring{P}} \mathring{N}$ is a smooth manifold. Now there is no difficulty in verifying that there exists a unique structure of a manifold with faces on $M \times_P N$ such that

(1) $\overset{\circ}{\overbrace{M \times_P N}} = \mathring{M} \times_{\mathring{P}} \mathring{N}$;

(2) the faces of $M \times_P N$ are of the form (a) $M_i \times_P N$ with collar $\tilde{F}_i$, $i \in I^{h,f}$; or (b) $M \times_P N_j$ with collar $\tilde{G}_j$, $j \in I^{h,g}$; or (c) $M_i \times_{P_i} N_i$ with collar H_i, $i \in I^{v,f} \cap I^{v,g}$.

4.3.7. Let M and N be manifolds with faces and $f : M \longrightarrow N$ be a smooth map such that for any face M_i of M, $f(M_i)$ is contained in a face N_i of N (however, f is not assumed to be compatible with the faces).

Let $f_i : M_i \longrightarrow N_i$ be the restriction of f. An easy verification shows that if f is submersive, then f_i is also submersive, $i \in I_M$.

4.3.8. Let M be a manifold with faces and $\{U_\alpha\}$ be an open covering of M. Then there exists a partition of unity $\{\psi_\alpha\}$ subordinated to the covering $\{U_\alpha\}$ and consisting of smooth mappings $\psi_\alpha : M \longrightarrow R$ compatible with the faces.

The proof is simple and left to the reader. The idea is to construct first the ψ_α's on the M_i's (by induction on dim(M)), then to extend them on a small neighborhood of each M_i (by composing with p_{M_i}) and finally to construct the ψ_α's on the whole of M by using the fact that $\overset{\circ}{M}$ is a smooth manifold without boundary (in this case the assertion is well known).

4.4. Let M be a manifold with faces and ξ be a smooth vector field on an open subset U of M. We shall call ξ parallel to the faces (denoted $\xi \in \mathcal{X}_{M,bM}(U)$) if, for any $i \in I_M$ such that $M_i \cap U \neq \phi$,

$$dp_{M_i} \cdot \xi(x) = \xi(p_{M_i}(x)) \quad \text{near } M_i \cap U \tag{4.4.1}$$

and

$$dr_{M_i} \cdot \xi = 0 \quad \text{near } M_i \cap U \quad . \tag{4.4.2}$$

These conditions are equivalent to

$$dF_{M_i} \cdot \xi = (\xi | M_i \cap U) \times 0 \quad \text{near } M_i \cap U \quad . \tag{4.4.3}$$

If $\xi \in \mathcal{X}_{M,bM}(U)$ then its flow $\lambda_\xi : D_\xi \longrightarrow U$ has the usual properties (i.e., D_ξ is open in $U \times R$, λ_ξ is smooth and for $u \in U$, if $(\{u\} \times R) \in D_\xi = \{u\} \times (s_u^\xi, t_u^\xi)$, then $t \longmapsto \lambda_\xi(u, t) : (s_u^\xi, t_u^\xi) \longrightarrow U$ is the maximal integral curve of ξ through u (the same definition as in 2.2)). Moreover λ_ξ is compatible with the faces (D_ξ and U have canonical structures of manifolds with faces).

4.5. LEMMA. Let M and N be manifolds with faces and $f : M \longrightarrow N$ be a smooth submersion compatible with the faces. Let $\eta \in X_{N,bN}(N)$, let $I \subset I_M$ and for any $i \in I$ let $\xi_i \in X_{M_i,bM_i}(M)$. Assume that $df \cdot \xi_i = \eta$ and $\xi_i | M_i \cap M_j = \xi_j | M_i \cap M_j$, $i, j \in I$. Then there exists $\xi \in X_{M,bM}(M)$ such that $df \cdot \xi = \eta$ and $\xi | M_i = \xi_i$, $i \in I$. If in addition f is proper then $D_\xi = (f \times 1_R)^{-1}(D_\eta)$.

Since the proof of this lemma is standard and raises no serious difficulty, it is left to the reader.

5. ABSTRACT STRATIFICATIONS (WITH FACES)

5.1. We shall now enlarge the notion of (w.)a.s. by allowing the strata to be manifolds with faces. Normally these objects should be called (weak) abstract stratifications with faces. However, in order to keep the terminology as simple as possible, we shall call them (weak) abstract stratifications and call the objects introduced in 1.2.1 w.a.s.'s without boundary. Now the definition.

A weak abstract stratification (w.a.s.) $\underline{\underline{A}}$ consists of (i) a nice topological space A; (ii) a locally finite family $\mathcal{A}$ of locally closed subsets of A (the strata) such that A is the disjoint union of the strata; (iii) a family of tubes of the strata $\{\tau_X = (T_X, \pi_X, \rho_X); X \in \mathcal{A}\}$; (iv) a family of closed subsets A_i $(i \in I_{\underline{\underline{A}}})$ of A, called the faces. The strata, their tubes and the faces must satisfy the following axioms:

(5.1.1) if $X, Y \in \mathcal{A}$ and $X \cap c\ell_A(Y) \neq \phi$, then $X \subset Y$;

(5.1.2) for any face A_i of $\underline{\underline{A}}$ there exist an open neighborhood U_{A_i} of A_i in A and a homeomorphism $F_{A_i} : U_{A_i} \longrightarrow A_i \times R_+$ onto an open subset of $A_i \times R_+$ such that (1) $F_{A_i}(a) = (a, 0)$, $a \in A_i$ and (2) for any $X \in \mathcal{A}$ $F_{A_i}(X \cap U_{A_i}) \subset (X \cap A_i) \times R_+$; F_{A_i} is called the collar of A_i; we shall define $p_{A_i} : U_{A_i} \longrightarrow A_i$ and $r_{A_i} : U_{A_i} \longrightarrow R_+$ by $F_{A_i}(a) = (p_{A_i}(a), r_{A_i}(a))$;

(5.1.3) each stratum $X \in A$ is a manifold with faces (in the induced topology), its faces being $X_i = X \cap A_i$, $i \in I_{\underline{\underline{A}}}$; the collar of X_i is $F_{X_i} = F_{A_i}|X \cap U_{A_i} : U_{X_i} = X \cap U_{A_i} \longrightarrow X_i \times R_+$;

(5.1.4) if $X \in A$ and $i \in I_{\underline{\underline{A}}}$, then $\pi_X^{-1}(X_i) = A_i \cap T_X$ and $F_{X_i} \circ \pi_X = ((\pi_X|T_X \cap A_i) \times 1_{R_+}) \circ F_{A_i}$ near X_i;

(5.1.5) for any $X \in A$ there exists ε_X such that, for any stratum $Y \neq X$ of $\underline{\underline{A}}$, $T_X^{\varepsilon_X} \cap Y \neq \phi$ implies that $X < Y$ and $(\pi_X, \rho_X)|T_X^{\varepsilon_X} \cap Y : T_X^{\varepsilon_X} \cap Y \to X \times (0, \varepsilon_X)$ is smooth and submersive;

(5.1.6) for any strata $X < Y$ the tubes τ_X and τ_Y verify (1.1.4) with $\varepsilon = \varepsilon_X$ and $\delta = \varepsilon_Y$.

If in addition

(5.1.7) for any $X \in A$ and $i \in I_{\underline{\underline{A}}}$, $\rho_X = \rho_X \circ p_{A_i}$ near X_i

and

(5.1.8) for any strata $X < Y$ the tubes τ_X and τ_Y verify (1.1.5) with $\varepsilon = \varepsilon_X$ and $\delta = \varepsilon_Y$,

then $\underline{\underline{A}}$ is called an <u>abstract stratification</u> (a.s.).

Let $\underline{\underline{A}}$ be a w.a.s. Since A is a normal space we can assume without any loss of generality that

(5.1.9) for any strata X and Y, $T_X^{\varepsilon_X} \cap T_Y^{\varepsilon_Y} \neq \phi$ if and only if $X \leq Y$ or $Y < X$.

Remark. Let $\underline{\underline{A}}$ be a w.a.s. and $X < Y$ be strata. By (5.1.4) $\pi_X | T_X^{\varepsilon_X} \cap Y : T_X^{\varepsilon_X} \cap Y \longrightarrow X$ is compatible with the faces, all the faces of $T_X^{\varepsilon_X} \cap Y$ being $(\pi_X | T_X^{\varepsilon_X} \cap Y)$-vertical. If $\underline{\underline{A}}$ is an a.s., by (5.1.7), $\rho_X | T_X^{\varepsilon_X} \cap Y : T_X^{\varepsilon_X} \cap Y \longrightarrow R$ is also compatible with the faces, all the faces of $T_X^{\varepsilon_X} \cap Y$ being $(\rho_X | T_X^{\varepsilon_X} \cap Y)$-horizontal.

Given a w.a.s. $\underline{\underline{A}}$ and $X \in \mathcal{A}$ we define $\operatorname{depth}_{\underline{\underline{A}}}(X)$, $\operatorname{depth}(\underline{\underline{A}})$ and $\dim(\underline{\underline{A}})$ exactly as in 1.2.2.

For formal reasons the empty set will be considered an a.s. of depth -1.

5.2. Let $\underline{\underline{A}}$ and $\underline{\underline{B}}$ be w.a.s.'s. A continuous map $f : B \longrightarrow A$ is called <u>compatible with the faces</u> if $I_{\underline{\underline{B}}} = I^{v,f} \cup I^{h,f}$, $I^{v,f} \subset I_{\underline{\underline{A}}}$ and

(5.2.1) if $i \in I^{v,f}$ then $f^{-1}(A_i) = B_i$ and

$F_{A_i} \circ f = ((f|B_i) \times 1_{R_+}) \circ F_{B_i}$ near B_i;

(5.2.2) if $i \in I^{h,f}$ then $f = (f|B_i) \circ p_{B_i}$ near B_i;

(5.2.3) if $i \in I_{\underline{\underline{A}}} \setminus I^{v,f}$ then $f^{-1}(A_i) = \phi$ or $f^{-1}(A_i) = B$.

The faces B_i of $\underline{\underline{B}}$ with $i \in I^{v,f}$ (resp. $i \in I^{h,f}$) are called f-<u>vertical</u> (resp. f-<u>horizontal</u>). If $i \in I^{h,f}$ we set $A_i = A$. For any $i \in I_{\underline{\underline{B}}}$ we define $f_i : B_i \to A_i$ to be the restriction of f.

If $f : B \longrightarrow A$ is compatible with the faces and

(5.2.4) for any $Y \in \mathcal{B}$ there exists $X \in \mathcal{A}$ such that $f(Y) \subset X$ and $f|Y : Y \longrightarrow X$ is smooth;

(5.2.5) Y and X being so above, the tubes τ_Y and τ_X satisfy (1.1.6),

then f is called a <u>weak morphism</u> (denoted $f : \underline{\underline{B}} \dashrightarrow \underline{\underline{A}}$).

If $f : \underline{\underline{B}} \dashrightarrow \underline{\underline{A}}$ and

(5.2.6) for any X and Y as in (5.2.4), the tubes τ_Y and τ_X satisfy (1.1.7)

then f is called a <u>morphism</u> (denoted $f : \underline{\underline{B}} \longrightarrow \underline{\underline{A}}$).

Notice that for a weak morphism $f : \underline{\underline{B}} \dashrightarrow \underline{\underline{A}}$ and two strata Y and X as in (5.2.4), the restriction $f|Y : Y \longrightarrow X$ is compatible with the faces.

A weak morphism $f : \underline{\underline{B}} \dashrightarrow \underline{\underline{A}}$ is called <u>submersive</u> if for any $Y \in \mathcal{B}$ and $X \in \mathcal{A}$ with $f(Y) \subset X$, $f|Y : Y \longrightarrow X$ is a submersion.

A (weak) morphism f is called a <u>(weak) isomorphism</u> if f^{-1} exists and is a (weak) morphism.

Two w.a.s.'s $\underline{\underline{A}}$ and $\underline{\underline{A}}'$ are called <u>equal</u> if $A = A'$ and 1_A is an isomorphism of $\underline{\underline{A}}$ on $\underline{\underline{A}}'$. If 1_A is only a weak isomorphism of $\underline{\underline{A}}$ on $\underline{\underline{A}}'$ then $\underline{\underline{A}}$ and $\underline{\underline{A}}'$ are called <u>weakly equal</u>.

5.3. Examples, remarks and constructions.

5.3.1. Let $\underline{\underline{A}}$ be a (w.)a.s. and $W \subset A$ be a locally closed subset. Set $\mathcal{A}|W = \{X \cap W;\ X \in \mathcal{A}$ and $X \cap W \neq \phi\}$. For any $i \in I_{\underline{\underline{A}}}$ set $W_i = A_i \cap W$. Suppose that for any $X \cap W \in \mathcal{A}|W$ there is given a subset $T_{X \cap W} \subset W \cap \pi_X^{-1}(X \cap W)$ containing $X \cap W$. We can define $\pi_{X \cap W} : T_{X \cap W} \longrightarrow X \cap W$ and $\rho_{X \cap W} : T_{X \cap W} \longrightarrow R_+$ to be the restrictions of π_X and ρ_X respectively. Set $\tau_{X \cap W} = (T_{X \cap W}, \pi_{X \cap W}, \rho_{X \cap W})$. If

(5.3.1.1) each $T_{X \cap W}$ is open in W;

(5.3.1.2) for any $i \in I_{\underline{\underline{A}}}$ either $W_i = \phi$ or $W_i = W$ or $W = (p_{A_i})^{-1}(W_i)$ near W_i (thus, in the last case, there exists an open neighborhood U_{W_i} of W_i in W such that $F_{W_i} = F_{A_i}|U_{W_i}$ is a homeomorphism of U_{W_i} onto an open subset of $W_i \times R_+$);

(5.3.1.3) if $X \cap W$, $Y \cap W \in \mathcal{A}|W$ and $(X \cap W) \cap c\ell_W(Y \cap W) \neq \phi$ then $X \cap W \subset c\ell_W(Y \cap W)$;

(5.3.1.4) each $X \cap W \in \mathcal{A}|W$ is a submanifold of X;

(5.3.1.5) any $X \cap W \in A|W$ verifies (5.1.5) (with X and Y replaced by $X \cap W$ and $Y \cap W$ respectively)

then one can endow W with a (w.)a.s. structure, denoted $\underline{\underline{A}}|W$ and called the restriction of $\underline{\underline{A}}$ to W, such that the inclusion $W \subset A$ determines a morphism $\underline{\underline{A}}|W \longrightarrow \underline{\underline{A}}$. The strata of $\underline{\underline{A}}|W$ are of the form $X \cap W$ with $X \cap W \in A|W$, their tubes being of the form $\tau_{X \cap W}$; the faces of $\underline{\underline{A}}|W$ are of the form W_i with $i \in I_{\underline{\underline{A}}}$ and $W_i \neq W$, the collar of W_i being F_{W_i} (see (5.3.1.2)).

As in the case of w.a.s.'s without boundary it follows that for any $X \cap W \in A|W$

$$T_{X \cap W} = W \cap \pi_X^{-1}(X \cap W) \quad \text{near} \quad X \cap W$$

(see (1.2.3.5)). Thus $\underline{\underline{A}}|W$, if it exists, is completely determined by $\underline{\underline{A}}$ and W.

The most important examples of subsets $W \subset A$ for which $\underline{\underline{A}}|W$ exists are provided by locally closed subsets which are union of strata or by open subsets.

If $\underline{\underline{A}}|W$ exists and for any $X \cap W \in A|W$

$$\pi_X^{-1}(X \cap W) = W \quad \text{near} \quad X \quad , \tag{5.3.1.6}$$

then W is called saturated.

For any $i \in I_{\underline{\underline{A}}}$, A_i is a saturated subset of A; in particular $\underline{\underline{A}}|A_i$ exists (in order to verify (5.3.1.5), one has to apply 4.3.7). From now on, unless something else is specified, we shall denote $\underline{\underline{A}}|A_i$ by $\underline{\underline{A}}_i$. Let $bA = \bigcup_{i \in I_{\underline{\underline{A}}}} A_i$ be the boundary of $\underline{\underline{A}}$ and $\mathring{A} = A \setminus bA$. Then $\mathring{A}$ is open in A and thus we can consider $\mathring{\underline{\underline{A}}} = \underline{\underline{A}}|\mathring{A}$. Clearly $\mathring{\underline{\underline{A}}}$ is a (w.)a.s. without boundary.

5.3.2. A manifold with faces M has a unique a.s. structure $\underline{\underline{M}}$ such that $M = \{M\}$ and the faces of $\underline{\underline{M}}$ are those of M. R_+ will be always considered a manifold with faces with only one face $\{0\}$, its collar being the obvious map $R_+ \longrightarrow \{0\} \times R_+$.

5.3.3. Let $\underline{\underline{A}}$ and $\underline{\underline{B}}$ be w.a.s.'s. Then $A \times B$ can be endowed with an obvious w.a.s. structure, denoted $\underline{\underline{A}} \times \underline{\underline{B}}$ (the strata and the tubes are defined as in 1.2.9 while the faces are defined as in 4.3.3). $\underline{\underline{A}} \times \underline{\underline{B}}$ is an a.s. if and only if $\underline{\underline{A}}$ and $\underline{\underline{B}}$ are a.s. and $\mathrm{depth}(\underline{\underline{A}}) \cdot \mathrm{depth}(\underline{\underline{B}}) = 0$.

We shall describe now a related and useful construction. Let $\underline{\underline{A}}$ be a (w.)a.s. and let $\varepsilon, \delta : \underline{\underline{A}} \longrightarrow \underline{\underline{R}}$, $\varepsilon < \delta$. Define $F_\varepsilon : A \times [\varepsilon, \delta) \longrightarrow (A \times \{\varepsilon\}) \times R_+$ and $F_\delta : A \times (\varepsilon, \delta] \longrightarrow (A \times \{\delta\}) \times R_+$ by $F_\varepsilon(a, t) = (a, \varepsilon(a), t - \varepsilon(a))$ and $F_\delta(a, t) = (a, \delta(a), \delta(a) - t)$. There exists an obvious (w.)a.s. structure on $A \times [\varepsilon, \delta]$, denoted $\underline{\underline{A}} \times [\varepsilon, \delta]$, such that the strata are of the form $X \times [\varepsilon|X, \delta|X]$ with $X \in \mathcal{A}$ and the faces are of the form (1) $A_i \times [\varepsilon|A_i, \delta|A_i]$ with $i \in I_{\underline{\underline{A}}}$ (its collar being a suitable restriction of $F_{A_i} \times 1_{R_+}$) or (2) $A \times \{\varepsilon\}$ (its collar being F_ε), or (3) $A \times \{\delta\}$ (its collar being F_δ). The fact that ε and δ are weak morphisms is important for this construction. In a similar way one can define $\underline{\underline{A}} \times \{\varepsilon\}$, $\underline{\underline{A}} \times (\varepsilon, \delta]$, $\underline{\underline{A}} \times [\varepsilon, \delta)$ and $\underline{\underline{A}} \times (\varepsilon, \delta)$ (this last one is simply $\underline{\underline{A}} \times \underline{\underline{R}}|A \times (\varepsilon, \delta)$).

5.3.4. Let $\underline{\underline{A}}$ be a (w.)a.s. and let A_i be a face of $\underline{\underline{A}}$. A direct verification shows that (possibly after shrinking U_{A_i}) F_{A_i} is a (weak) isomorphism of $\underline{\underline{A}}|U_{A_i}$ on $(\underline{\underline{A}}_i \times \underline{\underline{R}}_+)|F_{A_i}(U_{A_i})$ (here $\underline{\underline{A}}_i = \underline{\underline{A}}|A_i$, see (5.3.1). From now on we shall always choose U_{A_i} such that this assertion is true.

5.3.5. Let $\underline{\underline{A}}$ be a w.a.s. and M be a manifold with faces. As in 1.2.8 we can introduce the sheaf $C^\infty_{\underline{\underline{A}}}(\cdot, M)$ of controlled maps defined on open subsets of A and with values in M (in this new setting a controlled map is also compatible with the faces). If $M = R$ we shall denote this sheaf by $C^\infty_{\underline{\underline{A}}}$; its sections (over any subset of A) are called <u>controlled functions</u>. Lemma 1.3 and Corollary 1.4 are valid in this more general setting too (the proof of Lemma 1.3 can be carried out without any change; one has only to take into account 4.3.8).

5.3.6. Let $\underline{\underline{B}}$ and $\underline{\underline{A}}$ be w.a.s.'s and $f : B \longrightarrow A$ be continuous and compatible with the faces. We shall say that f is a quasi-morphism, denoted $f : \underline{\underline{B}} \rightsquigarrow \underline{\underline{A}}$, if

(1) for any $Y \in \mathcal{B}$, $f(Y)$ is a union of strata of $\underline{\underline{A}}$;

(2) for any $Y \in \mathcal{B}$ and $X \in \mathcal{A}$ with $X \subset c\ell_A(f(Y))$ there exist ε and δ such that $b \in T_Y^\delta \cap f^{-1}(T_X^\varepsilon)$ implies $\pi_Y(b) \in f^{-1}(T_X)$ and $\pi_X(f(b)) = \pi_X(f(\pi_Y(b)))$;

(3) for any Y and X as above $\pi_X \circ f|Y \cap f^{-1}(T_X^\varepsilon)$: $Y \cap f^{-1}(T_X^\varepsilon) \longrightarrow X$ is a smooth submersion and $\rho_X \circ f|Y \cap f^{-1}(T_X^\varepsilon)$: $Y \cap f^{-1}(T_X^\varepsilon) \longrightarrow R$ is smooth.

Clearly a weak morphism $f : \underline{\underline{B}} \dashrightarrow \underline{\underline{A}}$ is a quasi-morphism if and only if it is submersive and, for any $Y \in \mathcal{B}$, $f(Y) \in \mathcal{A}$.

Let now $f : \underline{\underline{B}} \rightsquigarrow \underline{\underline{A}}$ be a proper quasi-morphism, all the faces of $\underline{\underline{B}}$ being f-vertical. Let $C(f)$ be the topological mapping cylinder of f, i.e., $C(f)$ is obtained from the disjoint sum $(B \times [0, 1]) \sqcup (A \times \{0\})$ by identifying $(b, 0)$ with $(f(b), 0)$, $b \in B$. The image in $C(f)$ of an element $(c, t) \in (B \times [0, 1) \sqcup (A \times \{0\})$ is denoted $[c, t]$. Let $\bar{f} : B \sqcup A \longrightarrow A$, $\pi_f : C(p) \longrightarrow A$, $i_f : A \longrightarrow C(f)$, $j_f : B \longrightarrow C(f)$, $\tilde{A}$, $\tilde{B}$ and $\varphi_f : B \times [0, 1] \longrightarrow C(f)$ be given by $\bar{f}|B = f$, $\bar{f}|A = 1_A$, $\pi_f([c, t]) = \bar{f}(c)$, $i_f(a) = [a, 0]$, $j_f(b) = [b, 1]$, $\tilde{A} = i_f(A)$, $\tilde{B} = j_f(B)$, $\varphi_f(b, t) = [b, t]$.

Given a face B_i of $\underline{\underline{B}}$, let A_i be the corresponding face of $\underline{\underline{A}}$ (i.e., $B_i = f^{-1}(A_i)$) and set $f_i = f|B_i : B_i \longrightarrow A_i$. Then we can define $\dot{C}(f_i)$ and consider it as a subset of $C(f)$, more precisely $C(f_i) = (\pi_f)^{-1}(A_i)$. Since f is proper we may assume that $U_{B_i} = f^{-1}(U_{A_i})$ and the equality in (5.2.1) holds on U_{B_i}. Set $U_{C(f_i)} = \pi_f^{-1}(U_{A_i})$ and define $F_{C(f_i)} : U_{C(f_i)} \longrightarrow C(f_i) \times R_+$ by

$$F_{C(f_i)}([c,\ t]) = \begin{cases} ([p_{B_i}(c),\ t],\ r_{B_i}(c)),\ c \in U_{B_i},\ 0 \le t \le 1 \\ ([p_{A_i}(c),\ 0],\ r_{A_i}(c)),\ c \in U_{A_i},\ t = 0 \end{cases} .$$

Given $j \in I_{\underline{\underline{A}}} \smallsetminus I^{v,f}$ set $\tilde{A}_j = i_f(A_j)$, $U_{\tilde{A}_j} = i_f(U_{A_j})$ and define $F_{\tilde{A}_j} : U_{\tilde{A}_j} \longrightarrow \tilde{A}_j \times R_+$ by

$$F_{\tilde{A}_j}(i_f(a)) = (i_f(p_{A_j}(a)),\ r_{A_j}(a))$$

(notice that condition (3) in the definition of a quasi-morphism implies that

$f^{-1}(A_j) = \phi$).

Let $U_{\tilde{B}} = C(f) \smallsetminus \tilde{A}$ and define $F_{\tilde{B}} : U_{\tilde{B}} \longrightarrow \tilde{B} \times R_+$ by

$$F_{\tilde{B}}([b,\ t]) = ([b,\ 1],\ 1\text{-}t) \quad .$$

Let $\beta_1 : R \times R \longrightarrow R$, $\beta_2 : R \times R \times R \longrightarrow R$ and $\alpha : R_+ \times R_+ \longrightarrow R$ be given by

$$\beta_1(s,\ t) = \begin{cases} 0,\ t\text{-}s \le 0 \\ e^{1/s-t},\ s\text{-}t < 0 \end{cases}$$

$$\beta_2(s,\ t,\ u) = \beta_1(s,\ u)\beta_1(u,\ t)$$

$$\alpha(x,\ y) = \begin{cases} \dfrac{\int_y^{2x} \beta_2(x,\ 2x,\ u)du}{\int_x^{2x} \beta_2(x,\ 2x,\ u)du},\ (x,\ y) \ne (0,\ 0) \\ 0,\ (x,\ y) = (0,\ 0) \end{cases} .$$

Then α is smooth on $R_+ \times R_+ \smallsetminus \{0,\ 0\}$, $\alpha(x,\ y) = 0$ iff $y \ge 2x$, $\alpha(x,\ y) = 1$ iff $y \le x$ and $(x,\ y) \ne (0,\ 0)$ and $d\alpha_{(x,y)} \ne 0$ iff $x < y < 2x$.

Finally let $\varepsilon : A \longrightarrow R_+^*$ be a controlled function. A straightforward verification, which we omit, shows that there exists a w.a.s. structure

structure being given by the obvious identification of X with $\tilde{X}$; $T_{\tilde{X}} = \pi_f^{-1}(T_X)$, $\pi_{\tilde{X}}([c, t]) = [\pi_X(\bar{f}(c)), 0]$, $\rho_{\tilde{X}}([c, t]) =$ $= \alpha(t, \rho_X(\bar{f}(c)))\cdot\varepsilon(\pi_X(\bar{f}(c)))\cdot t + (1 - \alpha(t, \rho_X(\bar{f}(c))))\cdot\rho_X(\bar{f}(c))$;

(ii) $\varphi_f | B \times (0, 1] : \underline{\underline{B}} \times (0, 1] \longrightarrow \underline{\underline{C}}(f)^\varepsilon | C(f) \setminus \tilde{A}$ is an isomorphism;

(iii) the faces of $\underline{\underline{C}}(f)^\varepsilon$ are $\tilde{B}$, $C(f_i)$ $(i \in I_{\underline{\underline{B}}} = I^{v,f})$ and $\tilde{A}_j (j \in I_{\underline{\underline{A}}} \setminus I_{\underline{\underline{B}}})$ with collars $F_{\tilde{B}}$, $F_{C(f_i)}$ and $F_{\tilde{A}_j}$ respectively;

(iv) $\underline{\underline{C}}(f)^\varepsilon$ is uniquely determined by (i)-(iii);

(v) $i_f : \underline{\underline{A}} \longrightarrow \underline{\underline{C}}(f)^\varepsilon | \tilde{A}$ and $j_f : \underline{\underline{B}} \longrightarrow \underline{\underline{C}}(f)^\varepsilon | \tilde{B}$ are isomorphisms;

(vi) $\pi_f : \underline{\underline{C}}(f)^\varepsilon \rightsquigarrow \underline{\underline{A}}$ is a proper, surjective quasi-morphism, $\tilde{B}$ being the unique π_f-horizontal face of $\underline{\underline{C}}(f)^\varepsilon$; if in addition f is a weak morphism, then π_f is a weak morphism too;

(vii) let $i \in I_{\underline{\underline{B}}}$; then on $C(f_i)$ there are two w.a.s. structures: one inherited from $C(f)^\varepsilon$ (i.e., $\underline{\underline{C}}(f)^\varepsilon | C(f_i)$, cf. 5.3.1), the other one being $\underline{\underline{C}}(f_i)^{\varepsilon | A_i}$ (since $f_i : \underline{\underline{B}}_i \rightsquigarrow \underline{\underline{A}}_i$ is a proper quasimorphism, all the faces of $\underline{\underline{B}}_i$ being f_i-vertical, this makes sense); it is not difficult to see that they are equal;

(viii) let $X < X'$ be strata of $\underline{\underline{A}}$ such that the tubes τ_X and $\tau_{X'}$ verify the control condition (1.1.5). Then the tubes $\tau_{\tilde{X}}$ and $\tau_{\tilde{X}'}$ of $\underline{\underline{C}}(f)^\varepsilon$ verify (1.1.5) too. This is the only point where the so complicated definition of $\rho_{\tilde{X}}$ is needed (otherwise we could have defined $\rho_{\tilde{X}}$ by the formula $\rho_{\tilde{X}}([c, t]) = \varepsilon(\pi_X(\bar{f}(c)))\cdot t + \rho_X(\bar{f}(c))$;

(ix) if $\underline{\underline{B}}$ is an a.s. and $\text{depth}(\underline{\underline{A}}) = 0$, then $\underline{\underline{C}}(f)^\varepsilon$ is an a.s. (in this case the definition of $\rho_{\tilde{X}}$ in (i) above reduces to $\rho_{\tilde{X}}([c, t]) =$ $= \varepsilon(\pi_X(\bar{f}(c)))\cdot t)$.

Finally a remark concerning the notation: if $\varepsilon(a) = 1$, $a \in A$, then $\underline{\underline{C}}(f)^\varepsilon$ will be denoted $\underline{\underline{C}}(f)$.

5.3.7. Let $\underline{\underline{A}}$ (resp. $\underline{\underline{B}}$) be a (w.)a.s., A_0 (resp. B_0) be a face of $\underline{\underline{A}}$ (resp. $\underline{\underline{B}}$) and $f : \underline{\underline{A}}_0 \longrightarrow \underline{\underline{B}}_0$ be an isomorphism. Let $A \cup_f B$ be the topological space obtained from the disjoint sum of A and B by identifying $a \in A_0$ with $f(a) \in B_0$. We shall consider A and B as subspaces of $A \cup_f B$. Set $W = U_{A_0} \cup U_{B_0} \subset A \cup_f B$ and define $f : W \longrightarrow A_0 \times R$ by

$$F(a) = (p_{A_0}(a), -r_{A_0}(a)), \quad a \in U_{A_0} \quad ,$$
$$F(b) = (f^{-1}(p_{B_0}(b)), r_{B_0}(b)), \quad b \in U_{B_0} \quad .$$

Let $X \in \mathcal{A}$ and $Y \in \mathcal{B}$ and suppose that $f(X \cap A_0) = Y \cap B_0 \neq \phi$. Let $f_X = f|X \cap A_0 : X \cap A_0 \longrightarrow Y \cap B_0$ and set $Z = X \cup_{f_X} Y$ (see 4.3.4). Consider also $X' \in \mathcal{A}$ and $Y' \in \mathcal{B}$ such that $X < Y'$, $Y < Y'$ and $f(X' \cap A_0) = Y' \cap B_0$. As above define $f_{X'}$ and $Z' = X' \cup_{f_{X'}} Y'$. Choose $\varepsilon : X \longrightarrow R_+^*$ and $\delta : Y \longrightarrow R_+^*$ such that $\varepsilon|X \cap A_0 = \delta \circ f_X$. Then $T_Z = T_X^\varepsilon \cup T_Y^\delta$ is open in $A \cup_f B$ and, if ε and δ are sufficiently small, we can define $\rho_Z : T_Z \longrightarrow R_+$ by $\rho_Z|T_X^\varepsilon = \rho_X^\varepsilon$ and $\rho_Z|T_Y^\delta = \rho_Y^\delta$ (this is possible since f is an isomorphism). Assume now that for any X, X', Y and Y' as above

(5.3.7.1) $\rho_Z|T_Z \cap Z' : T_Z \cap Z' \longrightarrow R_+^*$ is smooth

(if $\underline{\underline{A}}$ and $\underline{\underline{B}}$ are a.s.'s this condition is always verified; it is a consequence of (5.1.7)).

A direct verification shows now that there exists a unique (w.)a.s. structure $\underline{\underline{A}} \cup_f \underline{\underline{B}}$ on $A \cup_f B$ such that

(i) the strata of $\underline{\underline{A}} \cup_f \underline{\underline{B}}$ are of the form $X \in \mathcal{A}$ (resp. $Y \in \mathcal{B}$) with $X \cap A_0 = \phi$ (resp. $Y \cap B_0 = \phi$) or $X \cup_{f_X} Y$ with $X \in \mathcal{A}$, $Y \in \mathcal{B}$ and $f(X \cap A_0) = Y \cap B_0 \neq \phi$;

(ii) $(\underline{\underline{A}} \cup_f \underline{\underline{B}})|A \setminus A_0 = \underline{\underline{A}}|A \setminus A_0$;

(iii) $(\underline{\underline{A}} \cup_f \underline{\underline{B}})|B \setminus B_0 = \underline{\underline{B}}|B \setminus B_0$;

(iv) $F : (\underline{\underline{A}} \cup_f \underline{\underline{B}})|W \dashrightarrow \underline{\underline{A}}_0 \times \underline{\underline{R}}$ is a (weak) isomorphism near A_0;

(v) for strata of $\underline{\underline{A}} \cup_f \underline{\underline{B}}$ of the form $Z = X \cup_{f_X} Y$, T_Z and ρ_Z are defined as above.

Notice that if $\underline{\underline{A}}$ and $\underline{\underline{B}}$ are a.s.'s, then (v) is a consequence of (iv).

If $A_0 = B_0$ and $f = 1_{A_0}$, $\underline{\underline{A}} \cup_f \underline{\underline{B}}$ will be denoted $\underline{\underline{A}} \cup_{A_0} \underline{\underline{B}}$. If $A_0 = \phi = B_0$, $\underline{\underline{A}} \cup_f \underline{\underline{B}}$ will be denoted $\underline{\underline{A}} \sqcup \underline{\underline{B}}$; thus $\underline{\underline{A}} \sqcup \underline{\underline{B}}$ is the (w.)a.s. whose underlying topological space is the disjoint sum of A and B, and $(\underline{\underline{A}} \sqcup \underline{\underline{B}})|A = \underline{\underline{A}}$ and $(\underline{\underline{A}} \sqcup \underline{\underline{B}})|B = \underline{\underline{B}}$.

Let $\underline{\underline{A}}$, $\underline{\underline{B}}$ and $\underline{\underline{C}}$ be w.a.s.'s, A_0 (resp. B_0, B_1, C_0) be a face of $\underline{\underline{A}}$ (resp. $\underline{\underline{B}}$, $\underline{\underline{B}}$, $\underline{\underline{C}}$) and $f : \underline{\underline{A}}_0 \longrightarrow \underline{\underline{B}}_0$ and $g : \underline{\underline{B}}_1 \longrightarrow \underline{\underline{C}}_0$ be isomorphisms which verify (5.3.7.1). We can consider $\underline{\underline{A}} \cup_f \underline{\underline{B}}$ and $\underline{\underline{B}} \cup_g \underline{\underline{C}}$. Assume that $B_0 \cap B_1 = \phi$. Then B_1 (resp. B_0) is a face of $\underline{\underline{A}} \cup_f \underline{\underline{B}}$ (resp. $\underline{\underline{B}} \cup_g \underline{\underline{C}}$) and we can consider $(\underline{\underline{A}} \cup_f \underline{\underline{B}}) \cup_g \underline{\underline{C}}$ and $\underline{\underline{A}} \cup_f (\underline{\underline{B}} \cup_g \underline{\underline{C}})$. It is obvious that we can identify $(A \cup_f B) \cup_g C$ with $A \cup_f (B \cup_g C)$ and then $(\underline{\underline{A}} \cup_f \underline{\underline{B}}) \cup_g \underline{\underline{C}}$ and $\underline{\underline{A}} \cup_f (\underline{\underline{B}} \cup_g \underline{\underline{C}})$ are equal. We shall denote this w.a.s. by $\underline{\underline{A}} \cup_f \underline{\underline{B}} \cup_g \underline{\underline{C}}$.

5.3.8. Let $q' : \underline{\underline{B}}' \rightsquigarrow \underline{\underline{A}}'$ and $q'' : \underline{\underline{B}}'' \rightsquigarrow \underline{\underline{A}}''$ be proper quasi-morphisms, all the faces of $\underline{\underline{B}}'$ (resp. $\underline{\underline{B}}''$) being q' (resp. q'')-vertical. Let B'_0 (resp. B''_0) be a face of $\underline{\underline{B}}'$ (resp. $\underline{\underline{B}}''$), let A'_0 (resp. A''_0) be the corresponding face of $\underline{\underline{A}}'$ (resp. $\underline{\underline{A}}''$) and let $q'_0 : \underline{\underline{B}}'_0 \rightsquigarrow \underline{\underline{A}}'_0$ (resp. $q''_0 : \underline{\underline{B}}''_0 \rightsquigarrow \underline{\underline{A}}''_0$) be the restriction of q' (resp. q''). Let $f : \underline{\underline{B}}'_0 \longrightarrow \underline{\underline{B}}''_0$ and $g : \underline{\underline{A}}'_0 \longrightarrow \underline{\underline{A}}''_0$ be isomorphisms such that $g \circ q'_0 = q''_0 \circ f$. Assume we can consider $\underline{\underline{B}}' \cup_f \underline{\underline{B}}''$ and $\underline{\underline{A}}' \cup_g \underline{\underline{A}}''$ (i.e., condition (5.3.7.1) is verified in both cases). Consider also $\underline{\underline{C}}(q')^{\varepsilon'}$ and $\underline{\underline{C}}(q'')^{\varepsilon''}$ for some ε' and ε'' such that $\varepsilon'|A'_0 = (\varepsilon''|A''_0) \circ g$. Define $q : B' \cup_f B'' \longrightarrow A' \cup_g A''$ and $F : C(q'_0) \longrightarrow C(q''_0)$ by $q|B' = q'$, $q|B'' = q''$, $F([b, t]) = [f(b), t]$ if $b \in B'_0$, $0 \leq t \leq 1$ and $F([a, 0]) = [g(a), 0]$ if $a \in A'_0$. Then q is a proper quasi-morphism from $\underline{\underline{B}}' \cup_f \underline{\underline{B}}''$ to $\underline{\underline{A}}' \cup_g \underline{\underline{A}}''$, all the faces of $\underline{\underline{B}}' \cup_f \underline{\underline{B}}''$ are q-vertical and F is an isomorphism from $\underline{\underline{C}}(q'_0)^{\varepsilon'|A'_0}$ to $\underline{\underline{C}}(q''_0)^{\varepsilon''|A''_0}$. Let $\varepsilon : A' \cup_g A'' \longrightarrow R^*_+$ be given by $\varepsilon|A' = \varepsilon'$ and $\varepsilon|A'' = \varepsilon''$. We can consider $\underline{\underline{C}}(q)^{\varepsilon}$ and $\underline{\underline{C}}(q')^{\varepsilon'} \cup_F \underline{\underline{C}}(q'')^{\varepsilon''}$ and define $\Phi : \underline{\underline{C}}(q')^{\varepsilon'} \cup_F \underline{\underline{C}}(q'')^{\varepsilon''} \longrightarrow \underline{\underline{C}}(q)^{\varepsilon}$ by $\Phi([b, t]) = [b, t]$(!). It is obvious that Φ is an isomorphism.

5.4. Let $\underline{\underline{A}}$ be a w.a.s. A vector field on $\underline{\underline{A}}$ is a family $\xi = \{\xi(x);\ \xi(x) \in TX_x,\ x \in X,\ X \in \mathcal{A}\}$. Given a vector field ξ on $\underline{\underline{A}}$ and a subset U of A, we define the restriction of ξ to U to be the family $\xi|U = \{\xi(x);\ x \in U\}$. The vector field ξ on $\underline{\underline{A}}$ is called weakly controlled if for any $X \in \mathcal{A}$ and any $i \in I_{\underline{\underline{A}}}$ there exist $\varepsilon < \varepsilon_X$ (ε_X as in (5.1.5)) and a neighborhood V_i of A_i in U_{A_i} such that

(5.4.1) $\quad \xi|X$ is a smooth vector field on X;

(5.4.2) $\quad$ for any $Y \in \mathcal{A}$ and any $y \in T_X^{\varepsilon} \cap Y$, $d\pi_X \cdot \xi(y) = \xi(\pi_X(y))$

and

(5.4.3) $\quad$ for any $a \in V_i$, $dF_{A_i} \cdot \xi(a) = (\xi(p_{A_i}(a)), 0)$.

If in addition, the notation being as in (5.4.2),

(5.4.4) $$d\rho_X \cdot \xi(y) = 0$$

ξ is called a controlled vector field on $\underline{\underline{A}}$.

As in 2.1 we can define the sheaf $\mathcal{X}^w_{\underline{\underline{A}}}$ of weakly controlled vector fields on open subsets of A and its subsheaf $\mathcal{X}_{\underline{\underline{A}}}$ of controlled vector fields. Both are sheaves of Lie algebras and $C^\infty_{\underline{\underline{A}}}$-modules. Also any $\xi \in \mathcal{X}^w_{\underline{\underline{A}}}(U)$ (U open in A) can be viewed as a derivation of $C^\infty_{\underline{\underline{A}}}(U)$ (the same formula as in 2.1).

Given an open subset $U \subset A$ and $\xi \in \mathcal{X}^w_{\underline{\underline{A}}}(U)$ we can define, exactly as in 2.2, the flow associated to ξ and denote it $\lambda_\xi : D_\xi \longrightarrow U$ (the construction is possible since, in view of (5.4.1) and (5.4.3), for any stratum X of $\underline{\underline{A}}|U$ the restriction $\xi|X$ is a smooth vector field parallel to the faces and therefore the remarks in 4.4 can be applied).

Let now $i \in I_{\underline{\underline{A}}}$. By the convention established in 5.3.4, $F_{\underline{\underline{A}}_i} : U_{A_i} \longrightarrow A_i \times R_+$ is a weak isomorphism of $\underline{\underline{A}}|U_{A_i}$ on $(\underline{\underline{A}}_i \times \underline{\underline{R}}_+)|F_{A_i}(U_{A_i})$. Let η_{A_i} be the unique vector field on U_{A_i} such that $dF_{A_i} \cdot \eta_{A_i} = 0 \times d/dt$. Clearly F_{A_i} is determined by η_{A_i}. Since $0 \times d/dt$ is controlled on $\underline{\underline{A}}_i \times \underline{\underline{R}}^*_+$ (but not on $\underline{\underline{A}}_i \times \underline{\underline{R}}_+$!) it follows that

$\eta_{A_i}|U_{A_i} \setminus A_i \in \mathcal{X}^w_{\underline{\underline{A}}}(U_{A_i} \setminus A_i)$. If $\underline{\underline{A}}$ is an a.s. then F_{A_i} is an isomorphism and therefore $\eta_{A_i}|U_{A_i} \setminus A_i \in \mathcal{X}_{\underline{\underline{A}}}(U_{A_i} \setminus A_i)$.

5.5 Let $\underline{\underline{B}}$ be a w.a.s. and $\underline{\underline{A}}$ be an a.s. A weak morphism $f : \underline{\underline{B}} \dashrightarrow \underline{\underline{A}}$ is called an <u>abstract Thom mapping</u> (a.T.m.), denoted $f : \underline{\underline{B}} \vdash\!\!\longrightarrow \underline{\underline{A}}$ if

(5.5.1) it is submersive;

(5.5.2) for any strata $Y < Y'$ of $\underline{\underline{B}}$, such that $f(Y)$ and $f(Y')$ are contained in the same stratum of $\underline{\underline{A}}$, the tubes τ_Y and $\tau_{Y'}$ verify (1.1.5);

(5.5.3) for any strata X of $\underline{\underline{A}}$ and Y of $\underline{\underline{B}}$ with $f(Y) \subset X$ and any $i \in I_{\underline{\underline{B}}}$, $\rho_Y = \rho_Y \circ p_{B_i}$ near $Y \cap B_i$ in $f^{-1}(X)$;

(5.5.4) for any strata $Y < Y'$ of $\underline{\underline{B}}$ with $f(Y) \subset X \in \mathcal{A}$ and $f(Y') \subset X' \in \mathcal{A}$ there exist $\delta < \varepsilon_Y$ and $\varepsilon < \varepsilon_X$ (ε_Y and ε_X as in (5.1.5)) such that $f(T^\delta_Y) \subset T^\varepsilon_X$ and the diagram

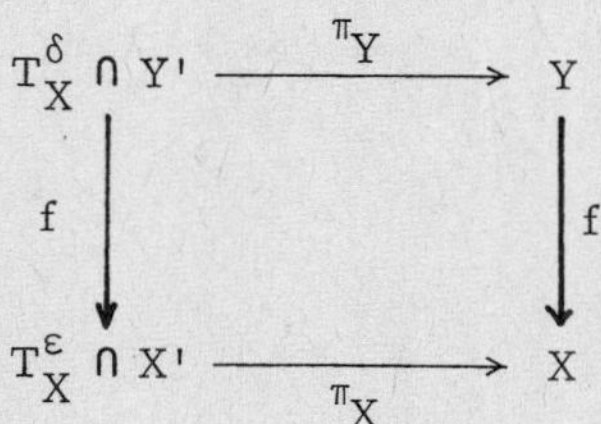

is regular.

<u>Remarks</u>: (1) Let $f : \underline{\underline{B}} \dashrightarrow \underline{\underline{A}}$ be a weak morphism, $\underline{\underline{A}}$ being an a.s. Assume that f is locally an a.T.m., i.e., B can be covered by open subsets U such that $f|U : \underline{\underline{B}}|U \dashrightarrow \underline{\underline{A}}$ is an a.T.m. Then f itself is an a.T.m.

(2) Let $f : \underline{\underline{B}} \vdash\!\!\longrightarrow \underline{\underline{A}}$ be an a.T.m. and $V \subset B$ (resp. $U \subset A$) be either a locally closed subset which is a union of strata or an open subset. Assume that $f(V) \subset U$. Then $f|V : \underline{\underline{B}}|V \dashrightarrow \underline{\underline{A}}|U$ is an a.T.m.

(3) Let $f : \underline{\underline{B}} \vdash\!\longrightarrow \underline{\underline{A}}$ be an a.T.m. and B_i be a face of $\underline{\underline{B}}$. Let $f_i : \underline{\underline{B}}_i \dashrightarrow \underline{\underline{A}}_i$ be the restriction of f (recall that $\underline{\underline{A}}_i = \underline{\underline{A}}$ if $i \in I^{h,f}$). Then f_i is an a.T.m.

The notion of morphism (resp. isomorphism) between a.T.m.'s is defined exactly as in 3.2.

Let $f : \underline{\underline{B}} \vdash\!\longrightarrow \underline{\underline{A}}$ be an a.T.m. and $V \subset B$ be an open subset. We shall denote by $X^f_{\underline{\underline{B}}}(V)$ the set of those vector fields $\eta \in X^w_{\underline{\underline{B}}}(V)$ which verify

(5.5.5) for any $X \in A$, $\eta | f^{-1}(X) \cap V$ is a controlled vector field on $\underline{\underline{B}} | f^{-1}(X) \cap V$.

It is clear that the collection $\{X^f_{\underline{\underline{B}}}(V);\ V$ open in $B\}$ is a subsheaf of $X^w_{\underline{\underline{B}}}$; we shall denote it $X^f_{\underline{\underline{B}}}$. In fact $X^f_{\underline{\underline{B}}}$ is a $C^\infty_{\underline{\underline{B}}}$-submodule and also a subsheaf of Lie algebras of $X^w_{\underline{\underline{B}}}$.

Taking into account 4.4, 4.5 and 5.3.5, the following proposition can be proved exactly as Lemma 2.3, Lemma 3.5 and Lemma 3.6.

5.6. PROPOSITION. Let $f : \underline{\underline{B}} \vdash\!\longrightarrow \underline{\underline{A}}$ be an a.T.m., $B' \subset B$ be a closed union of strata and let $I \subset I_{\underline{\underline{B}}}$. Let $f' : B' \longrightarrow A$ and $f_i : B_i \longrightarrow A_i$, $i \in I$, be the restrictions of f (recall that $A_i = A$ if $i \in I^{h,f}$). Set $\underline{\underline{B}}' = \underline{\underline{B}} | B'$. Let $\xi \in X^w_{\underline{\underline{A}}}(A)$, $\eta' \in X^{f'}_{\underline{\underline{B}}'}(B')$ and $\eta_i \in X^{f_i}_{\underline{\underline{B}}_i}(B_i)$ $(i \in I)$ be such that

(1) $\eta' | B' \cap B_i = \eta_i | B' \cap B_i$, $i \in I$;

(2) $\eta_i | B_i \cap B_j = \eta_j | B_i \cap B_j$, $i, j \in I$;

(3) $df \cdot \eta' = \xi$;

(4) $df \cdot \eta_i = \xi$, $i \in I$.

Then

(i) there exists $\eta \in X^f_{\underline{\underline{B}}}(B)$ such that $df \cdot \eta = \xi$, $\eta | B' = \eta'$ and $\eta | B_i = \eta_i$, $i \in I$;

(ii) if $\xi \in X_{\underline{\underline{A}}}(A)$, then D_ξ is open in $A \times R$, D_η is open in $B \times R$ and the pair $(\lambda_\eta, \lambda_\xi)$ is a morphism from

$(f \times 1_R)|D_\eta : (\underline{\underline{B}} \times \underline{\underline{R}})|D_\eta \longmapsto (\underline{\underline{A}} \times \underline{\underline{R}})|D_\xi$ to $f : \underline{\underline{B}} \longmapsto \underline{\underline{A}}$; if in addition f is proper, then $D_\eta = (f \times 1_R)^{-1}(D_\xi)$.

5.7. THEOREM (Thom second isotopy lemma). Let $f : \underline{\underline{B}} \longmapsto \underline{\underline{A}}$ be a proper a.T.m., M be a manifold with faces and $p : A \longrightarrow M$ be a proper controlled submersion. Then f is locally trivial over p (same definition as in 3.3.3).

Proof. The assertion being local with respect to M, and f and p being proper and compatible with the faces, we may assume that

(i) $M = R^m \times (R_+)^q$ and $0 \in p(f(B))$;

(ii) $\underline{\underline{A}} = \underline{\underline{A}}' \times (\underline{\underline{R}}_+)^q$, $\underline{\underline{B}} = \underline{\underline{B}}' \times (\underline{\underline{R}}_+)^q$;

(iii) there exist a proper a.T.m. $f' : \underline{\underline{B}}' \longmapsto \underline{\underline{A}}'$ and a proper controlled submersion $p' : A' \longrightarrow R^m$ such that $f = f' \times 1_{(R_+)^q}$ and $p = p' \times 1_{(R_+)^q}$.

Clearly it is sufficient to prove that the a.T.m. f' is trivial over the proper controlled submersion p'. This can be done exactly as in the proof of Theorem 3.7. Q.E.D.

5.8. COROLLARY (Thom first isotopy lemma). Let $\underline{\underline{B}}$ and $\underline{\underline{A}}$ be a.s.'s and $f : \underline{\underline{B}} \longrightarrow \underline{\underline{A}}$ be proper and submersive. Suppose that $\text{depth}(\underline{\underline{A}}) = 0$. Then f is locally trivial.

Proof. It is obvious that f is an a.T.m. Since A is a manifold with faces and $1_A : A \longrightarrow A$ is a proper controlled submersion, Theorem 5.7 implies that f is trivial over 1_A, which means that f is locally trivial. Q.E.D.

5.9. We shall list now some results which can be derived exactly as the corresponding ones in the case of w.a.s.'s without boundary. When necessary we shall give some indications concerning their proof.

5.9.1. Let $\underline{\underline{A}}$ be an a.s. and $X < Y$ be strata of $\underline{\underline{A}}$. For a sufficiently small ε the map $(\pi_X, \rho_X)|T_X^\varepsilon \setminus Y : T_X^\varepsilon \setminus Y \longrightarrow X \times (0, \varepsilon)$ is surjective.

5.9.2. Let $\underline{\underline{B}}$ be an a.s., $\underline{\underline{A}}$ be a w.a.s., $A_0 \subset A$ be saturated (see 5.3.1) and $f : \underline{\underline{B}} \dashrightarrow \underline{\underline{A}}$ be transverse to A_0 (the same definition as in 2.9). Then $f^{-1}(A_0)$ is saturated in $\underline{\underline{B}}$; in particular $\underline{\underline{B}}|f^{-1}(A_0)$ exists. (The strata and the tubes of $\underline{\underline{B}}|f^{-1}(A_0)$ are defined as in 2.9. By 4.3.5 the strata are manifolds with faces. In order to check the conditions which must be verified by the strata and their tubes one proceeds as in 2.9. The faces are defined as in 4.3.5).

5.9.3. Let $\underline{\underline{B}}$ and $\underline{\underline{C}}$ be a.s.'s, $\underline{\underline{A}}$ be a w.a.s. and $f : \underline{\underline{B}} \dashrightarrow \underline{\underline{A}}$ and $g : \underline{\underline{C}} \dashrightarrow \underline{\underline{A}}$ be transverse, i.e., for any $Y \in \mathcal{B}$, $Z \in \mathcal{C}$ and $X \in \mathcal{A}$ with $f(Y) \cup g(Z) \subset X$, the smooth maps $f|Y : Y \longrightarrow X$ and $g|Z : Z \longrightarrow X$ are transverse. Set $B \times_A C = \{(b, c) \in B \times C;\ f(b) = g(c)\}$. Then $B \times_A C$ can be endowed with a canonical w.a.s. structure $\underline{\underline{B}} \times_A \underline{\underline{C}}$, called the <u>fibre product</u> of $\underline{\underline{B}}$ and $\underline{\underline{C}}$ over $\underline{\underline{A}}$ (with respect to f and g). If $\operatorname{depth}(\underline{\underline{B}}) \cdot \operatorname{depth}(\underline{\underline{C}}) = 0$, then $\underline{\underline{B}} \times_A \underline{\underline{C}}$ is an a.s.

(Since Δ_A is not saturated in $\underline{\underline{A}} \times \underline{\underline{A}}$ (if $bA \neq \phi$, then (5.3.1.2) does not hold) we cannot proceed as in 2.1.1. We shall proceed as follows. For any $Y \in \mathcal{B}$ and $Z \in \mathcal{C}$, by 4.3.6, $Y \times_A Z = (Y \times Z) \cap (B \times_A C)$ is a manifold with faces. Thus $\{Y \times_A Z;\ Y \in \mathcal{B},\ Z \in \mathcal{C}\}$ is a locally finite partition of $B \times_A C$ into locally closed subsets which are manifolds with faces. The faces of $\underline{\underline{B}} \times_A \underline{\underline{C}}$ are defined as in 4.3.6. The tube (T_W, π_W, ρ_W) of a stratum $W = Y \times_A Z$ is defined by:
$T_W = (T_Y^\varepsilon \times T_Z^\delta) \cap (B \times_A C)$ for some sufficiently small ε and δ,
$\pi_W(b, c) = (\pi_Y(b), \pi_Z(c))$ and $\rho_W(b, c) = \rho_Y(b) + \rho_Z(c)$, $(b, c) \in T_W$.
A straightforward verification completes the construction of $\underline{\underline{B}} \times_A \underline{\underline{C}}$).

5.9.4. If $\underline{\underline{B}}$ and $\underline{\underline{A}}$ are w.a.s.'s and $\operatorname{depth}(\underline{\underline{A}}) = 0$, then $f : \underline{\underline{B}} \dashrightarrow \underline{\underline{A}}$ is an a.T.m. if and only if $\underline{\underline{B}}$ is an a.s. and f is submersive.

5.9.5. Let $f : \underline{\underline{B}} \mapsto\!\!\!\dashrightarrow \underline{\underline{A}}$ be an a.T.m. and Y and X be strata of $\underline{\underline{B}}$ and $\underline{\underline{A}}$ respectively such that $f(Y) \subset X$. Choose ε and δ such that all the control conditions involved hold on T_X^ε and T_Y^δ and $f(T_Y^\delta) \subset T_X^\varepsilon$.

Consider the fibre product $\underline{\underline{Y}} \times_X (\underline{\underline{A}}|T_X^\varepsilon)$ of $\underline{\underline{Y}}$ and $\underline{\underline{A}}|T_X^\varepsilon$ over X (with respect to $f|Y$ and π_X^ε) and define $g : B_\varepsilon^\delta = T_Y^\delta \cap f^{-1}(T_X^\varepsilon) \longrightarrow Y \times_X T_X^\varepsilon$ by $g(b) = (\pi_Y(b), f(b))$. Then g is an a.T.m. from $\underline{\underline{B}}|B_\varepsilon^\delta$ to $\underline{\underline{Y}} \times_X (\underline{\underline{A}}|T_X^\varepsilon)$.

5.9.6. Let $f : \underline{\underline{B}} \vdash\!\!\longrightarrow \underline{\underline{A}}$ be an a.T.m. and $\underline{\underline{C}}$ be an a.s. of depth zero. Then $f \times 1_C : \underline{\underline{B}} \times \underline{\underline{C}} \dashrightarrow \underline{\underline{A}} \times \underline{\underline{C}}$ is an a.T.m.

5.9.7. Let $f' : \underline{\underline{B}}' \vdash\!\!\longrightarrow \underline{\underline{A}}'$ (resp. $f'' : \underline{\underline{B}}'' \vdash\!\!\longrightarrow \underline{\underline{A}}''$) be an a.T.m., B'_0 (resp. B''_0) be an f' (resp. f'')-vertical face of $\underline{\underline{B}}'$ (resp. $\underline{\underline{B}}''$) and A'_0 (resp. A''_0) be the corresponding face of $\underline{\underline{A}}'$ (resp. $\underline{\underline{A}}''$). Let (ψ, φ) be an isomorphism of $f'|B'_0 : \underline{\underline{B}}'_0 \vdash\!\!\longrightarrow \underline{\underline{A}}'_0$ on $f''|B''_0 : \underline{\underline{B}}''_0 \vdash\!\!\longrightarrow \underline{\underline{A}}''_0$ such that ψ verifies (5.3.7.1). We can therefore consider the w.a.s. $\underline{\underline{B}} = \underline{\underline{B}}' \cup_\psi \underline{\underline{B}}''$ and the a.s. $\underline{\underline{A}} = \underline{\underline{A}}' \cup_\varphi \underline{\underline{A}}''$. Define $f : B \longrightarrow A$ by setting $f|B' = f'$ and $f|B'' = f''$. A straightforward verification shows that $f : \underline{\underline{B}} \dashrightarrow \underline{\underline{A}}$ is an a.T.m. We shall denote f by $f' \cup_\psi f''$. If $\underline{\underline{B}}'_0 = \underline{\underline{B}}''_0$, $\underline{\underline{A}}'_0 = \underline{\underline{A}}''_0$, $\psi = 1_{B'_0}$ and $\varphi = 1_{A'_0}$ we shall denote f by $f' \cup_{B'_0} f''$.

5.9.8. Let $f' : \underline{\underline{B}}' \vdash\!\!\longrightarrow \underline{\underline{A}}$ (resp. $f'' : \underline{\underline{B}}'' \vdash\!\!\longrightarrow \underline{\underline{A}}$) be an a.T.m., B'_0 (resp. B''_0) be an f' (resp. f'')-horizontal face of $\underline{\underline{B}}'$ (resp. $\underline{\underline{B}}''$) and $(\psi, 1_A)$ be an isomorphism of $f'|B'_0 : \underline{\underline{B}}'_0 \vdash\!\!\longrightarrow \underline{\underline{A}}$ on $f''|\underline{\underline{B}}''_0 : \underline{\underline{B}}''_0 \vdash\!\!\longrightarrow \underline{\underline{A}}$ such that ψ verifies (5.3.7.1). Let $\underline{\underline{B}} = \underline{\underline{B}}' \cup_\psi \underline{\underline{B}}''$ and define $f : B \longrightarrow A$ by setting $f|B' = f'$ and $f|B'' = f''$. As above, a straightforward verification shows that $f : \underline{\underline{B}} \dashrightarrow \underline{\underline{A}}$ is an a.T.m. We shall denote it $f' \cup_\psi f''$; if $\underline{\underline{B}}'_0 = \underline{\underline{B}}''_0$ and $\psi = 1_{B'_0}$ we shall use the notation $f' \cup_{B'_0} f''$.

5.9.9. Let $f : \underline{\underline{C}} \vdash\!\!\longrightarrow \underline{\underline{B}}$ and $g : \underline{\underline{B}} \vdash\!\!\longrightarrow \underline{\underline{A}}$ be a.T.m.'s In general $g \circ f$ is not an a.T.m. However if for any $X \in \mathcal{A}$ the strata of $\underline{\underline{B}}$ contained in $g^{-1}(X)$ are not comparable (i.e., if $Y_i \in \mathcal{B}$ and $g(Y_i) \in X$, $i = 1, 2$, then $Y_1 \leq Y_2$ implies $Y_1 = Y_2$) then $g \circ f : \underline{\underline{C}} \dashrightarrow \underline{\underline{A}}$ is an a.T.m.

(Indeed, (5.5.1) is obvious, (5.5.2) and (5.5.3) are easy consequences of our additional assumption, and (5.5.4) follows from 3.1.2).

5.10. PROPOSITION. Let $f : \underline{\underline{B}} \vdash\!\longrightarrow \underline{\underline{A}}$ be an a.T.m., W be an open subset of A and $g : W \longrightarrow R$ be continuous. Suppose that there exist $t \in R$ and an open neighborhood U of $A^t = g^{-1}(t)$ in W such that $A^{t]} = g^{-1}((-\infty, t])$ is closed in A and $g|U$ is a controlled submersion. Set $A^{t)} = A^{t]} \smallsetminus A^t$, $A^{[t} = A \smallsetminus A^{(t}$, $A^{(t} = A^{[t} \smallsetminus A^t$, $B^t = f^{-1}(A^t)$, $B^{[t} = f^{-1}(A^{[t})$ and similarly $B^{(t}$, $B^{t]}$ and $B^{t)}$. Define $f^t : B^t \longrightarrow A^t$, $f^{t]} : B^{t]} \longrightarrow A^{t]}$ and $f^{[t} : B^{[t} \longrightarrow A^{[t}$ to be the restrictions of f. Then $B^{[t}$ and $B^{t]}$ (resp. $A^{[t}$ and $A^{t]}$) can be endowed with w.a.s. structures $\underline{\underline{B}}^{[t}$ and $\underline{\underline{B}}^{t]}$ (resp. a.s. structures $\underline{\underline{A}}^{[t}$ and $\underline{\underline{A}}^{t]}$) with the following properties:

(i) $\underline{\underline{B}}^{[t}|B^{(t} = \underline{\underline{B}}|B^{(t}$, $\underline{\underline{B}}^{t]}|B^{t)} = \underline{\underline{B}}|B^{t)}$, $\underline{\underline{A}}^{[t}|A^{(t} = \underline{\underline{A}}|A^{(t}$ and $\underline{\underline{A}}^{t]}|A^{t)} = \underline{\underline{A}}|A^{t)}$ (notice that $A^{t)}$ and $A^{(t}$ are open in A and $B^{t)}$ and $B^{(t}$ are open in B);

(ii) $f^{t]}$ (resp. $f^{t]}$) is an a.T.m. from $\underline{\underline{B}}^{t]}$ (resp. $\underline{\underline{B}}^{[t}$) to $\underline{\underline{A}}^{t]}$ (resp. $\underline{\underline{A}}^{[t}$);

(iii) A^t is a face of $\underline{\underline{A}}^{[t}$ (resp. $\underline{\underline{A}}^{t]}$); the restriction of $\underline{\underline{A}}$ to A^t exists and $\underline{\underline{A}}|A^t = \underline{\underline{A}}^{[t}|A^t = \underline{\underline{A}}^{t]}|A^t$;

(iv) B^t is an $f^{t]}$ (resp. $f^{[t}$)-vertical face of $\underline{\underline{B}}^{t]}$ (resp. $B^{[t}$); the restriction of $\underline{\underline{B}}$ to B^t exists and $\underline{\underline{B}}|B^t = \underline{\underline{B}}^{t]}|B^t = \underline{\underline{B}}^{[t}|B^t$;

(v) one can construct $B^{t]} \cup_{B^t} B^{[t}$ and the inclusions $B^{t]} \subset B$, $B^{[t} \subset B$, $A^{t]} \subset A$ and $A^{[t} \subset A$ determine a weak isomorphism $G : \underline{\underline{B}}^{t]} \cup_{B^t} \underline{\underline{B}}^{[t} \dashrightarrow \underline{\underline{B}}$ and an isomorphism $F : \underline{\underline{A}}^{t]} \cup_{A^t} \underline{\underline{A}}^{[t} \longrightarrow \underline{\underline{A}}$ such that the pair (G, F) is an isomorphism from $f^{t]} \cup_{B^t} f^{[t}$ $: \underline{\underline{B}}^{t]} \cup_{B^t} B^{[t} \vdash\!\longrightarrow \underline{\underline{A}}^{t]} \cup_{A^t} \underline{\underline{A}}^{[t}$ to $f : \underline{\underline{B}} \vdash\!\longrightarrow \underline{\underline{A}}$;

(vi) assume that there exist $\xi \in \mathcal{X}_{\underline{\underline{A}}}(U)$ and $\eta \in \mathcal{X}^f_{\underline{\underline{B}}}(f^{-1}(U))$ such that $df \cdot \eta = \xi$ and $dg \cdot \xi = d/dt$; then we can choose $\underline{\underline{B}}^{t]}$ and $\underline{\underline{B}}^{[t}$ (resp. $\underline{\underline{A}}^{t]}$ and $\underline{\underline{A}}^{[t}$) such that the vector fields associated to the face B^t (resp. A^t) of $\underline{\underline{B}}^{t]}$ and $\underline{\underline{B}}^{[t}$ (resp. $\underline{\underline{A}}^{t]}$ and $\underline{\underline{A}}^{[t}$) (see 5.4) are restrictions of η (resp. ξ).

<u>Proof</u>. Let $V = f^{-1}(U)$. Then $f|V : \underline{\underline{B}}|V \vdash\!\longrightarrow \underline{\underline{A}}|U$ and $g|U : \underline{\underline{A}}|U \vdash\!\longrightarrow \underline{\underline{R}}$ are a.T.m.'s and by Proposition 5.6 there exist

$\xi \in \mathcal{X}_{\underline{\underline{A}}}(U)$ and $\eta \in \mathcal{X}^f_{\underline{\underline{B}}}(V)$ such that $dg \cdot \xi = d/dt$ and $df \cdot \eta = \xi$. If $\varepsilon : A^t \longrightarrow R^*_+$ and $\delta : B^t \longrightarrow R^*_+$ are sufficiently small, then $A^t \times (-\varepsilon, \varepsilon) \subset D_\xi$ and $B^t \times (-\delta, \delta) \subset D_\eta$. Set $U_\varepsilon = \lambda_\xi(A^t \times (-\varepsilon, \varepsilon))$ and $V_\delta = \lambda_\eta(B^t \times (-\delta, \delta))$. Clearly

$$U_\varepsilon = \{a \in U;\ (a, -g(a)) \in D_\xi \text{ and } |g(a)| < \varepsilon(\lambda_\xi(a, -g(a)))\}$$

and

$$V_\delta = \{b \in V;\ (b, -g(f(b))) \in D_\eta \text{ and } |g(f(b))| < \delta(\lambda_\eta(b, -g(f(b))))\} .$$

Thus U_ε and V_δ are open in A and B respectively. Arguing as in the proof of Thom isotopy lemmas (Theorems 2.6 and 3.7) we can see that the restrictions of $\underline{\underline{B}}$ and $\underline{\underline{A}}$ to B^t and A^t respectively exist and f^t is an a.T.m. from $\underline{\underline{B}}^t = \underline{\underline{B}}|B^t$ to $\underline{\underline{A}}^t = \underline{\underline{A}}|A^t$. Moreover $F = \lambda_\xi|A^t \times (-\varepsilon, \varepsilon)$ is an isomorphism of $\underline{\underline{A}}^t \times (-\varepsilon, \varepsilon)$ on $\underline{\underline{A}}|U_\varepsilon$ and $G = \lambda_\eta|B^t \times (-\delta, \delta)$ is a weak isomorphism of $\underline{\underline{B}}^t \times (-\delta, \delta)$ on $\underline{\underline{B}}|V_\delta$. Notice that $F(A^t \times [0, \varepsilon)) = A^{[t} \cap U_\varepsilon$, $F(A^t \times (-\varepsilon, 0]) = A^{t]} \cap U_\varepsilon$, $G(B^t \times [0, \delta)) = B^{[t} \cap V_\delta$ and $G(B^t \times (-\delta, 0]) = B^{t]} \cap V_\delta$. It is now obvious that there exists a unique a.s. structure $\underline{\underline{A}}^{[t}$ on $A^{[t}$ such that $\underline{\underline{A}}^{[t}|A^{(t} = \underline{\underline{A}}|A^{(t}$ and A^t is a face of $\underline{\underline{A}}^{[t}$, its collar being $F^{-1}|A^{[t} \cap U_\varepsilon : A^{[t} \cap U_\varepsilon \longrightarrow A^t \times R_+$. One defines similarly $\underline{\underline{A}}^{t]}$, $\underline{\underline{B}}^{[t}$ and $\underline{\underline{B}}^{t]}$. The rest of the proof is a simple matter of verification. Q.E.D.

5.11. PROPOSITION. Let $f : \underline{\underline{B}} \longmapsto \underline{\underline{A}}$ be an a.T.m., $W \subset B$ be open and $g : W \longrightarrow R$ be continuous. Suppose that there exist $t \in R$, an open neighborhood V of $B^t = g^{-1}(t)$ in W and $\eta \in \mathcal{X}^f_B(V)$ such that

(1) g is controlled on V;

(2) $df \cdot \eta = 0$;

(3) $dg \cdot \eta = d/dt$;

(4) $B^{t]} = g^{-1}((-\infty, t])$ is closed in B.

Set $B^{t)} = B^{t]} \setminus B^t$, $B^{[t} = B \setminus B^{t)}$, $B^{(t} = B \setminus B^{t]}$ and define $f^t : B^t \longrightarrow A$, $f^{[t} : B^{[t} \longrightarrow A$ and $f^{t]} : B^{t]} \longrightarrow A$ to be the restrictions of f. Then there exist w.a.s. structures $\underline{\underline{B}}^{[t}$ and $\underline{\underline{B}}^{t]}$ on $B^{[t}$ and $B^{t]}$

respectively such that

(i) $\underline{\underline{B}}^{[t}|B^{(t} = \underline{\underline{B}}|B^{(t}$, $\underline{\underline{B}}^{t]}|B^{t)} = \underline{\underline{B}}|B^{t)}$;

(ii) $f^{[t}$ (resp. $f^{t]}$) is an a.T.m. from $\underline{\underline{B}}^{[t}$ (resp. $\underline{\underline{B}}^{t]}$) to $\underline{\underline{A}}$;

(iii) B^t is an $f^{[t}$ (resp. $f^{t]}$)-horizontal face of $\underline{\underline{B}}^{[t}$ (resp. $\underline{\underline{B}}^{t]}$) and the vector field associated to it (see 5.4) is a restriction of η; the restriction of $\underline{\underline{B}}$ to B^t exists and $\underline{\underline{B}}|B^t = \underline{\underline{B}}^{t]}|B^t = \underline{\underline{B}}^{[t}|B^t$;

(iv) one can construct $\underline{\underline{B}}^{t]} \cup_{B^t} \underline{\underline{B}}^{[t}$ and the inclusions $B^{t]} \subset B$ and $B^{[t} \subset B$ determine a weak isomorphism $G : \underline{\underline{B}}^{t]} \cup_{B^t} \underline{\underline{B}}^{[t} \dashrightarrow \underline{\underline{B}}$ such that $(G, 1_A)$ is an isomorphism from $f^{t]} \cup_{B^t} f^{[t} : \underline{\underline{B}}^{t]} \cup_{B^t} \underline{\underline{B}}^{[t} \longmapsto \underline{\underline{A}}$ to $f : \underline{\underline{B}} \longmapsto \underline{\underline{A}}$.

<u>Proof.</u> Similar to that of Proposition 5.10. Q.E.D.

5.12. Let $f : \underline{\underline{B}} \longmapsto \underline{\underline{A}}$ be an a.T.m. and let $X \in \mathcal{A}$ and $Y \in \mathcal{B}$ be such that $f(Y) \subset X$. If $\varepsilon : X \longrightarrow R_+^*$ is sufficiently small then ρ_X is controlled on $T_X^\varepsilon \setminus X$. In general this is not true for ρ_Y, i.e., ρ_Y is not controlled on $T_Y^\delta \setminus Y$, even if δ is very small. We shall prove however that ρ_Y is controlled near $(T_Y^\delta \setminus Y) \cap f^{-1}(X)$ if δ is sufficiently small.

For each $Z \in \mathcal{B}$ choose $\alpha(Z) : Z \longrightarrow R_+^*$ such that all the control conditions involved hold on $T_Z^{\alpha(Z)}$. Set $\delta = \alpha(Y)$ and $\mathcal{B}_Y^+ = \{Z \in \mathcal{B};\ Z > Y$ and $f(Z) \in X\}$. Then

(5.12.1) ρ_Y is controlled on $T_Y^\delta \cap (\bigcup_{Z \in \mathcal{B}_Y^+} T_Z^{\alpha(Z)})$.

Indeed, set $V = T_Y^\delta \cap (\bigcup_{Z \in \mathcal{B}_Y^+} T_Z^{\alpha(Z)})$ and notice that $V \cap Y = \phi$. Let $Z' \in \mathcal{B}$ be a stratum such that $Z' \cap V \neq \phi$ (thus $Z' \neq Y$) and let $b \in V \cap T_{Z'}^{\alpha(Z')}$. Then there exists $Z \in \mathcal{B}_Y^+$ such that $b \in T_Z^{\alpha(Z)}$. Since $T_Z^{\alpha(Z)} \cap T_{Z'}^{\alpha(Z')} \neq \phi$, $Z' \leq Z$ or $Z < Z'$. Also, since $T_Y^\delta \cap Z' \neq \phi$, $Y < Z'$. Let $X' \in \mathcal{A}$ be the stratum which contains $f(Z')$. Clearly $X \leq X'$. If $Z' \leq Z$, then $X = X'$ and by (5.5.2)

$$\rho_Y(\pi_{Z'}(b)) = \rho_Y(b).$$

If $Z < Z'$, then

$$\rho_Y(\pi_{Z'}(b)) = \rho_Y(\pi_Z(\pi_{Z'}(b))) = \rho_Y(\pi_Z(b)) = \rho_Y(b)$$

(the first and last equalities are consequences of (5.5.2), while the second one follows from (5.1.6)). Similar arguments show that $\rho_Y|V$ is compatible with the faces of $\underline{\underline{B}}|V$. Thus (5.12.1) is proved.

6. THE STRUCTURE OF ABSTRACT THOM MAPPINGS.

6.1. The aim of this chapter is to prove that any a.T.m. whose source and target have finite depth can be "decomposed" into "simpler" a.T.m. 's. As a first step we show how to "decompose" a.s. 's into "simpler" ones. What "decompose" and "simpler" mean will be explained in 6.1.1 for a.s.'s and in 6.8.2 for a.T.m.'s.

<u>Notation</u>. In all this chapter $\underline{\underline{A}}$ will denote a non empty a.s. of finite depth. Set $\mathcal{A}^* = \{X \in \mathcal{A};\ \text{depth}_{\underline{\underline{A}}}(X) = \text{depth}(\underline{\underline{A}})\}$ and $A^* = \bigcup_{X \in \mathcal{A}^*} X$. Any $X \in \mathcal{A}^*$ is closed in A and therefore A^* is also closed in A (as a locally finite union of closed subsets) and $\underline{\underline{A}}|A^*$ exists. Set $\underline{\underline{A}}^* = \underline{\underline{A}}|A^*$. Clearly $\text{depth}(\underline{\underline{A}}^*) = 0$ and $A^* \neq A$ if and only if $\text{depth}(\underline{\underline{A}}) > 0$.

Given $\varepsilon : \mathcal{A}^* \to R_+$ set $A^{\varepsilon)} = \bigcup_{X \in \mathcal{A}^*} T_X^{\varepsilon|X}$. If ε is sufficiently small, then $T_X^{\varepsilon|X} \cap T_Y^{\varepsilon|Y} = \emptyset$ for any strata $X \neq Y$ of $\underline{\underline{A}}^*$. Thus we can define $\pi^{\varepsilon)} : A^{\varepsilon)} \to A^*$ and $\rho^{\varepsilon)} : A^{\varepsilon)} \to R_+$ by setting

$$\pi^{\varepsilon)}|T_X^{\varepsilon|X} = \pi_X|T_X^{\varepsilon|X}, \quad \rho^{\varepsilon)}|T_X^{\varepsilon|X} = \rho_X|T_X^{\varepsilon|X}, \quad X \in \mathcal{A}^* .$$

The notation and the conventions introduced in 5.3.6 and 5.3.7 will be used without any other mention.

6.1.1. A quintuple $\Delta = \{\underline{\underline{A}}^-, \underline{\underline{A}}^+, p, \varepsilon, \Phi\}$ is called a <u>decomposition</u> of $\underline{\underline{A}}$ if

(i) $\underline{\underline{A}}^-$ and $\underline{\underline{A}}^+$ are a.s.'s, A^- and A^+ are closed subsets of A, $A^0 = A^- \cap A^+$ is a face of both $\underline{\underline{A}}^-$ and $\underline{\underline{A}}^+$, $\underline{\underline{A}}^-|A^0 = \underline{\underline{A}}^+|A^0$ and the inclusions $A^- \subset A$ and $A^+ \subset A$ determine an isomorphism of $\underline{\underline{A}}^- \cup_{A^0} \underline{\underline{A}}^+$ on $\underline{\underline{A}}$ (as a consequence $\underline{\underline{A}}|A^0$ exists and $\underline{\underline{A}}|A^0 = \underline{\underline{A}}^+|A^0 = \underline{\underline{A}}|A^0$; denote $\underline{\underline{A}}^0 = \underline{\underline{A}}|A^0$);

(ii) $p : \underline{\underline{A}}^0 \dashrightarrow \underline{\underline{A}}^*$ is a proper submersive weak morphism sending strata onto strata (i.e., if $X \in A^0$ then $p(X) \in A^*$) ;

(iii) $\varepsilon : \underline{\underline{A}}^* \longrightarrow \underline{\underline{R}}^*_+$

(iv) $\Phi : \underline{\underline{C}}(p)^\varepsilon \to \underline{\underline{A}}^-$ is an isomorphism such that

$$\Phi([a,0]) = a, \quad a \in A^*,$$

$$\Phi([a,1]) = a, \quad a \in A^0.$$

Remarks. (1) If depth($\underline{\underline{A}}$) = 0 and $\Delta = \{\underline{\underline{A}}^-, \underline{\underline{A}}^+, p, \varepsilon, \Phi\}$ is a decomposition of $\underline{\underline{A}}$, then $\underline{\underline{A}}^- = \underline{\underline{A}}^* = \underline{\underline{A}}$, $A^+ = A^0 = \emptyset$, $\underline{\underline{C}}(p)^\varepsilon = \underline{\underline{A}}$ and $\Phi = 1_A$.

(2) Let $\Delta = \{\underline{\underline{A}}^-, \underline{\underline{A}}^+, p, \varepsilon, \Phi\}$ be a decomposition of $\underline{\underline{A}}$. Then depth($\underline{\underline{A}}^+$) = depth($\underline{\underline{A}}$) - 1 and thus the structure of $\underline{\underline{A}}^+$ is simpler than that of $\underline{\underline{A}}$. On the other hand, although depth($\underline{\underline{A}}^-$) = depth($\underline{\underline{A}}$), the structure of $\underline{\underline{A}}^-$ is simpler than the structure of $\underline{\underline{A}}$: it is isomorphic to the structure of $\underline{\underline{C}}(p)^\varepsilon$ which is determined by $\underline{\underline{A}}^0$, $\underline{\underline{A}}^*$ and p, and depth ($\underline{\underline{A}}^0$) = depth($\underline{\underline{A}}$) - 1, depth($\underline{\underline{A}}^*$) = 0.

6.1.2. Let $\tilde{\underline{\underline{A}}}$ be another a.s. and $F : \underline{\underline{A}} \to \tilde{\underline{\underline{A}}}$ be an isomorphism. Let also $\Delta = \{\underline{\underline{A}}^-, \underline{\underline{A}}^+, p, \varepsilon, \Phi\}$ and $\tilde{\Delta} = \{\tilde{\underline{\underline{A}}}^-, \tilde{\underline{\underline{A}}}^+, \tilde{p}, \tilde{\varepsilon}, \tilde{\Phi}\}$ be decompositions of $\underline{\underline{A}}$ and $\tilde{\underline{\underline{A}}}$ respectively. F is called <u>compatible with</u> Δ <u>and</u> $\tilde{\Delta}$ if

(1) $F(A^-) = \tilde{A}^-$ and $F^- = F|A^- : \underline{\underline{A}}^- \to \tilde{\underline{\underline{A}}}^-$ is an isomorphism;

(2) $F(A^+) = \tilde{A}^+$ and $F^+ = F|A^+ : \underline{\underline{A}}^+ \to \tilde{\underline{\underline{A}}}^+$ is an isomorphism;

(3) clearly $F(A^0) = \tilde{A}^0$ and $F(A^*) = \tilde{A}^*$; let $F^0 = F|A^0 : A^0 \to \tilde{A}^0$ and $F^* = F|A^* : A^* \to \tilde{A}^*$; then $\tilde{p} \circ F^0 = F^* \circ p$ and $\varepsilon = \tilde{\varepsilon} \circ F^*$;

(4) $F(\Phi([a,t])) = \tilde{\Phi}([F(a),t]), \quad [a,t] \in C(p)$.

Two decompositions Δ and $\tilde{\Delta}$ of $\underline{\underline{A}}$ are called equal (denoted $\Delta = \tilde{\Delta}$) if 1_A is compatible with Δ and $\tilde{\Delta}$.

Let again $\tilde{\underline{\underline{A}}}$ be an a.s. and $F : \underline{\underline{A}} \to \tilde{\underline{\underline{A}}}$ be an isomorphism. Given a decomposition Δ of $\underline{\underline{A}}$, it is easy to see that there exists a unique decomposition $F_*(\Delta)$ of $\tilde{\underline{\underline{A}}}$ such that F is compatible with Δ and $F_*(\Delta)$.

6.1.3. Let $\Delta = \{\underline{\underline{A}}^-, \underline{\underline{A}}^+, p, \varepsilon, \Phi\}$ be a decomposition of $\underline{\underline{A}}$. Let $F^+_{A^0} : U^+_{A^0} \to A^0 \times R_+$ be the collar of A^0 in $\underline{\underline{A}}^+$, $p^+_{A^0} : U^+_{A^0} \to A^0$ and $r^+_{A^0} : U^+_{A^0} \to R_+$ be the associated maps and $\eta^+_{A^0}$ be the associated vector field. Set $U_\Delta = A^- \cup U^+_{A^0}$ and define $\pi_\Delta : U_\Delta \to A^*$, $\varphi_\Delta : U_\Delta \to R_+$ and $\xi_\Delta \in \mathcal{X}_{\underline{\underline{A}}}(U_\Delta \setminus A^*)$ as follows:

$$\pi_\Delta(\Phi([a,t])) = \pi_p([a,t]), \quad [a,t] \in C(p)$$

$$\pi_\Delta(a) = p(p^+_{A^0}(a)), \quad a \in U^+_{A^0}$$

$$\varphi_\Delta(\Phi([a,t])) = t, \quad [a,t] \in C(p)$$

$$\varphi_\Delta(a) = 1 + r^+_{A^0}(a), \quad a \in U^+_{A^0}$$

$$\xi_\Delta | A^- \setminus A^* = d(\Phi \circ \varphi_p) \cdot (0 \times d/dt)$$

$$\xi_\Delta | U^+_{A^0} \setminus A^0 = \eta^+_{A^0}$$

(in the above relations $\pi_p : C(p) \to A^*$ and $\varphi_p : A^0 \times [0,1] \to C(p)$ are the mappings introduced in 5.3.6). It is obvious from the definitions that

$$d\pi_\Delta \cdot \xi_\Delta = 0 \tag{6.1.3.1}$$

and

$$\varphi_\Delta | U_\Delta \setminus A^* \in C^\infty_{\underline{\underline{A}}}(U_\Delta \setminus A^*) \text{ and } d\varphi_\Delta \cdot \xi_\Delta = d/dt . \tag{6.1.3.2}$$

Also, since Φ is an isomorphism

(6.1.3.3) $$\pi_\Delta = \pi_X \quad \text{near} \quad X, \quad X \in A^*$$

and

(6.1.3.4) $$\varphi_\Delta = \rho_X/\varepsilon \circ \pi_X \quad \text{near} \quad X, \quad X \in A^* .$$

As in the case of tubes and collars we are only interested in the germ of U_Δ (resp. π_Δ, φ_Δ, ξ_Δ) at A^-. In order to maintain the notation as simple as possible, we shall always identify these germs with suitable chosen representatives.

Let now $\lambda_\Delta = \lambda_{\xi_\Delta} : D_\Delta = D_{\xi_\Delta} \to A$ be the flow associated to ξ_Δ. From the definitions it follows that

$$\{(a,t) \in (U_\Delta \smallsetminus A^*) \times R; \ -\varphi_\Delta(a) < t \leq 0\} = \{(a,t) \in D_\Delta; \ t \leq 0\};$$

$$\Phi([a,t]) = \lambda_\Delta(a,t-1), \quad a \in A^0, \quad 0 < t \leq 1 ;$$

$$\Phi([a,0]) = \lim_{t \nearrow 1} \lambda_\Delta(a,-t), \quad a \in A^0 ;$$

$$F^+_{A^0}(a) = (\lambda_\Delta(a, 1-\varphi_\Delta(a)), \ \varphi_\Delta(a)-1), \quad a \in U^+_{A^0} ;$$

$$A^* = \varphi_\Delta^{-1}(0), \quad A^- = \varphi_\Delta^{-1}([0,1]), \quad A^0 = \varphi_\Delta^{-1}(1) .$$

Given an open subset U of A and $\xi \in \mathcal{X}_{\underline{\underline{A}}}(U)$ set

$$U_\xi = \{a \in U; \ s^\xi_a \geq -1\} .$$

Then

$$U_{\xi_\Delta} = A^- \smallsetminus A^* .$$

The above relations imply that Δ is determined by ξ and ε. More precisely

6.1.4. LEMMA. Two decompositions Δ and $\tilde{\Delta}$ of $\underline{\underline{A}}$ are equal if and only if $\varepsilon = \tilde{\varepsilon}$, $U_{\xi_\Delta} = U_{\xi_{\tilde{\Delta}}}$ and $\xi_\Delta = \xi_{\tilde{\Delta}}$ near U_{ξ_Δ}.

6.1.5. Given a decomposition $\Delta = \{\underline{\underline{A}}^-, \underline{\underline{A}}^+, p, \varepsilon, \Phi\}$ of $\underline{\underline{A}}$ and a face A_i of $\underline{\underline{A}}$ such that $\operatorname{depth}(\underline{\underline{A}}_i) = \operatorname{depth}(\underline{\underline{A}})$ (this is equivalent to $(A_i)^* = A_i \cap A^*$ or to $A_i \cap A^* \neq \emptyset$), Δ induces in a canonical way a decomposition $\Delta|A_i = \{\underline{\underline{A}}_i^-, \underline{\underline{A}}_i^+, p_i, \varepsilon_i, \Phi_i\}$ of $\underline{\underline{A}}_i$ where $A_i^- = A^- \cap A_i$ and $A_i^+ = A^+ \cap A_i$ are faces of $\underline{\underline{A}}^-$ and $\underline{\underline{A}}^+$ respectively, $\underline{\underline{A}}_i^- = \underline{\underline{A}}^-|A_i$, $\underline{\underline{A}}_i^+ = \underline{\underline{A}}^+|A_i^+$, $A_i^0 = A_i^- \cap A_i^+ = A^0 \cap A_i$, p_i, ε_i and Φ_i are the restrictions of p, ε and Φ respectively. Note that

$$U_{\Delta|A_i} = U_\Delta \cap A_i$$

$$\varphi_{\Delta|A_i} = \varphi_\Delta|U_{\Delta|A_i}$$

$$\xi_{\Delta|A_i} = \xi_\Delta|U_{\Delta|A_i} .$$

6.1.6. Let $\Delta = \{\underline{\underline{A}}^-, \underline{\underline{A}}^+, p, \varepsilon, \Phi\}$ be a decomposition of $\underline{\underline{A}}$ and U be an open subset of A containing A^-. Then $\Delta|U = \{\underline{\underline{A}}^-, \underline{\underline{A}}^+|A^+ \cap U, p, \varepsilon, \Phi\}$ is a decomposition of $\underline{\underline{A}}|U$.

6.1.7. Let $\Delta = \{\underline{\underline{A}}^-, \underline{\underline{A}}^+, p, \varepsilon, \Phi\}$ be a decomposition of $\underline{\underline{A}}$ and let $\mu : \underline{\underline{A}}^* \dashrightarrow \underline{\underline{R}}|(0,1)$. Set $\tilde{A}^- = \{a \in A^-;\ \varphi_\Delta(a) \leq \mu(\pi_\Delta(a))\}$, $\tilde{A}^0 = \{a \in A^-;\ \varphi_\Delta(a) = \mu(\pi_\Delta(a))\}$ and $\tilde{A}^+ = A \setminus (\tilde{A}^- \setminus \tilde{A}^0)$. Define $\tilde{p} : \tilde{A}^0 \to A^*$, $\tilde{\varepsilon} : A^* \to R_+^*$ and $\tilde{\Phi} : C(\tilde{p}) \to \tilde{A}^-$ by $\tilde{p} = \pi_\Delta|\tilde{A}^0$, $\tilde{\varepsilon} = \mu\varepsilon$ and $\tilde{\Phi}([a,t]) = \Phi([a,\mu(\pi_\Delta(a))t])$. Then one can endow $\tilde{A}^-$ and $\tilde{A}^+$ with unique a.s. structures $\underline{\underline{\tilde{A}}}^-$ and $\underline{\underline{\tilde{A}}}^+$ such that $\Delta_\mu = \{\underline{\underline{\tilde{A}}}^-, \underline{\underline{\tilde{A}}}^+, \tilde{p}, \tilde{\varepsilon}, \tilde{\Phi}\}$ is a decomposition of $\underline{\underline{A}}$ and $U_{\tilde{\Delta}} = U_\Delta$, $\xi_{\Delta_\mu} = (\mu \circ \pi_\Delta)\xi_\Delta$, $\pi_{\Delta_\mu} = \pi_\Delta$ and $\varphi_{\Delta_\mu} = (1/\mu \circ \pi_\Delta)\,\varphi_\Delta$.

(Indeed, $\tilde{A}^- = \varphi_{\Delta_\mu}^{-1}((-\infty,1])$, $\tilde{A}^0 = \varphi_{\Delta_\mu}^{-1}(1)$ and $d\,\varphi_{\Delta_\mu} \cdot \xi = d/dt$. Proposition 5.10 gives us $\underline{\underline{\tilde{A}}}^-$ and $\underline{\underline{\tilde{A}}}^+$, the rest being a simple matter of verification.)

6.1.8. Suppose that $\underline{\underline{A}} = {}^1\underline{\underline{A}} \cup_B {}^2\underline{\underline{A}}$, B being a face of both ${}^1\underline{\underline{A}}$ and ${}^2\underline{\underline{A}}$, and ${}^1\underline{\underline{A}}|B = {}^2\underline{\underline{A}}|B$. Let ${}^i\Delta = \{{}^i\underline{\underline{A}}^-, {}^i\underline{\underline{A}}^+, {}^ip, {}^i\varepsilon, {}^i\Phi\}$ be a decomposition of ${}^i\underline{\underline{A}}$, $i = 1,2$. If $\operatorname{depth}({}^1\underline{\underline{A}}) = \operatorname{depth}({}^2\underline{\underline{A}}) = \operatorname{depth}(\underline{\underline{A}}|B)$, we assume that ${}^1\Delta|B = {}^2\Delta|B$. We shall define a decomposition ${}^1\Delta \cup_B {}^2\Delta = \{\underline{\underline{A}}^-, \underline{\underline{A}}^+, p, \varepsilon, \Phi\}$

of $\underline{\underline{A}}$ as follows.

<u>Case I</u>: $\text{depth}(\underline{\underline{A}}) = \text{depth}({}^1\underline{\underline{A}}) > \text{depth}({}^2\underline{\underline{A}})$. Then $A^* = {}^1A^*$, $B \cap {}^1A^- = \emptyset$ and B is a face of ${}^1\underline{\underline{A}}^+$. Take $\underline{\underline{A}}^- = {}^1\underline{\underline{A}}^-$, $\underline{\underline{A}}^+ = {}^1\underline{\underline{A}}^+ \cup_B {}^2\underline{\underline{A}}$, $p = {}^1p : A^0 = {}^1A^0 \to {}^1A^* = A^*$, $\varepsilon = {}^1\varepsilon$ and $\Phi = {}^1\Phi$.

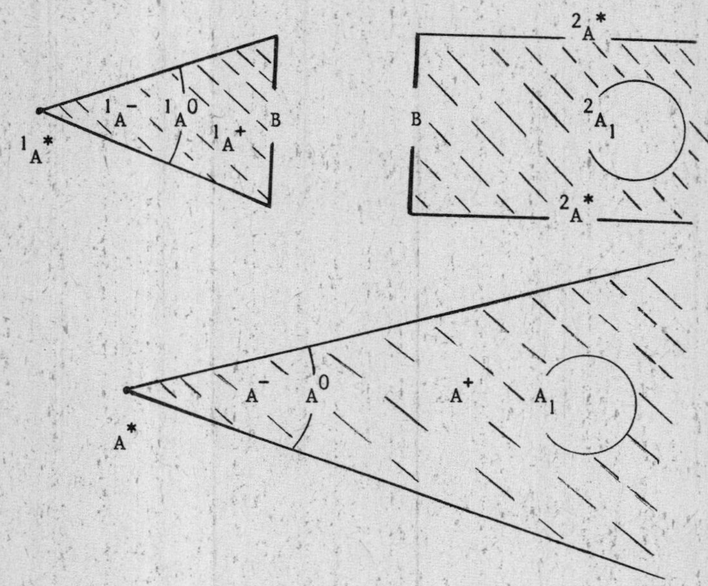

<u>Case II</u>: $\text{depth}(\underline{\underline{A}}) = \text{depth}({}^2\underline{\underline{A}}) > \text{depth}({}^1\underline{\underline{A}})$. This case is similar to Case I.

<u>Case III</u>: $\text{depth}(\underline{\underline{A}}) = \text{depth}({}^1\underline{\underline{A}}) = \text{depth}({}^2\underline{\underline{A}}) > \text{depth}(\underline{\underline{A}}|B)$. Then ${}^1A^- \cap {}^2A^- = \emptyset$, $A^* = {}^1A^* \sqcup {}^2A^*$, B is face of both ${}^1\underline{\underline{A}}^+$ and ${}^2\underline{\underline{A}}^+$ and ${}^1\underline{\underline{A}}^+|B = {}^1\underline{\underline{A}}|B = {}^2\underline{\underline{A}}|B = {}^2\underline{\underline{A}}^+|B$. Take $\underline{\underline{A}}^- = {}^1\underline{\underline{A}}^- \sqcup {}^2\underline{\underline{A}}^-$, ${}^+\underline{\underline{A}} = {}^1\underline{\underline{A}}^+ \cup_B {}^2\underline{\underline{A}}^+$; then $A^0 = A^- \cap A^+ = {}^1A^0 \sqcup {}^2A^0$. Define $p : A^0 \to A^*$ and $\varepsilon : A^* \to R^*_+$ by $p|{}^iA^0 = {}^ip$ and $\varepsilon|{}^iA^* = {}^i\varepsilon$, $i = 1,2$. Then $\underline{\underline{C}}(p)^{\varepsilon} = \underline{\underline{C}}({}^1p)^{{}^1\varepsilon} \sqcup \underline{\underline{C}}({}^2p)^{{}^2\varepsilon}$ and we define $\Phi : C(p) \to A^-$ by $\Phi|C({}^ip) = {}^i\Phi$, $i = 1,2$.

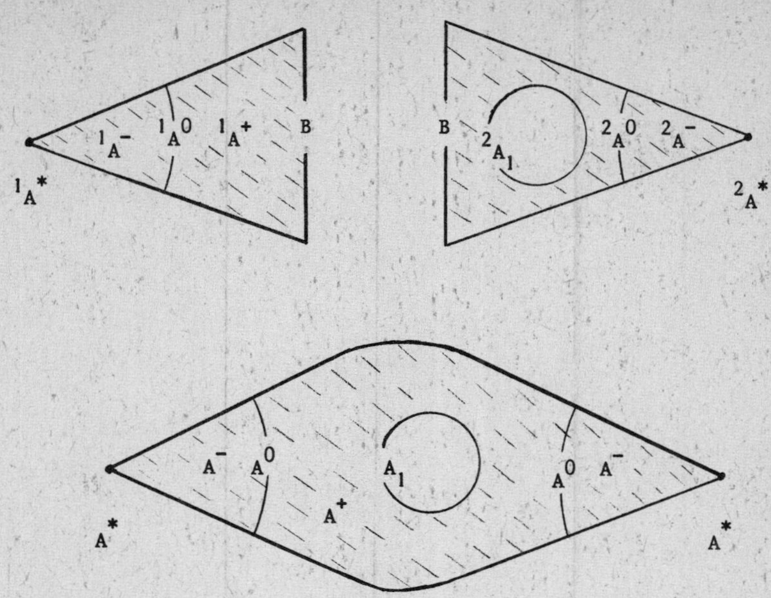

Case IV: $\operatorname{depth}(\underline{\underline{A}}) = \operatorname{depth}({}^1\underline{\underline{A}}) = \operatorname{depth}({}^2\underline{\underline{A}}) = \operatorname{depth}(\underline{\underline{A}}|B)$. Then, by hypothesis ${}^1\Delta|B = {}^2\Delta|B = \{\underline{\underline{B}}^-, \underline{\underline{B}}^+, q, \delta, \Psi\}$. Notice that

$$ {}^1A^* \cap {}^2A^* = B \cap {}^1A^* = B \cap {}^2A^* = B^* \ ; $$

$$ \underline{\underline{A}}^* = {}^1\underline{\underline{A}}^* \cup_{B^*} {}^2\underline{\underline{A}}^* \ ; $$

$$ {}^1A^- \cap {}^2A^- = B \cap {}^1A^- = B \cap {}^2A^- = B^- \ ; $$

$$ {}^1A^+ \cap {}^2A^+ = B \cap {}^1A^+ = B \cap {}^2A^+ = B^+ \ ; $$

$$ {}^1A^0 \cap {}^2A^0 = B \cap {}^1A^0 = B \cap {}^2A^0 = B^0 \ . $$

Take $\underline{\underline{A}}^- = {}^1\underline{\underline{A}}^- \cup_{B^-} {}^2\underline{\underline{A}}^-$ and $\underline{\underline{A}}^+ = {}^1\underline{\underline{A}}^+ \cup_{B^+} {}^2\underline{\underline{A}}^+$; then $A^0 = A^- \cap A^+ = {}^1A^0 \cup {}^2A^0$ and we can define $p : A^0 \to A^*$ and $\varepsilon : A^* \to R^*_+$ by $p|{}^iA^0 = {}^ip$, $\varepsilon|{}^iA^* = {}^i\varepsilon$, $i = 1,2$. Notice that $C(p) = C({}^1p) \cup C({}^2p)$ and $C({}^1p) \cap C({}^2p) = C(q)$. Thus we can define $\Phi : C(p) \to A^-$ by setting

$\Phi|C({}^i p) = {}^i\Phi$. As in the other cases, a straightforward verification shows that ${}^1\Delta \cup_B {}^2\Delta = \{\underline{\underline{A}}^-, \underline{\underline{A}}^+, p, \varepsilon, \Phi\}$ is a well defined decomposition of $\underline{\underline{A}}$.

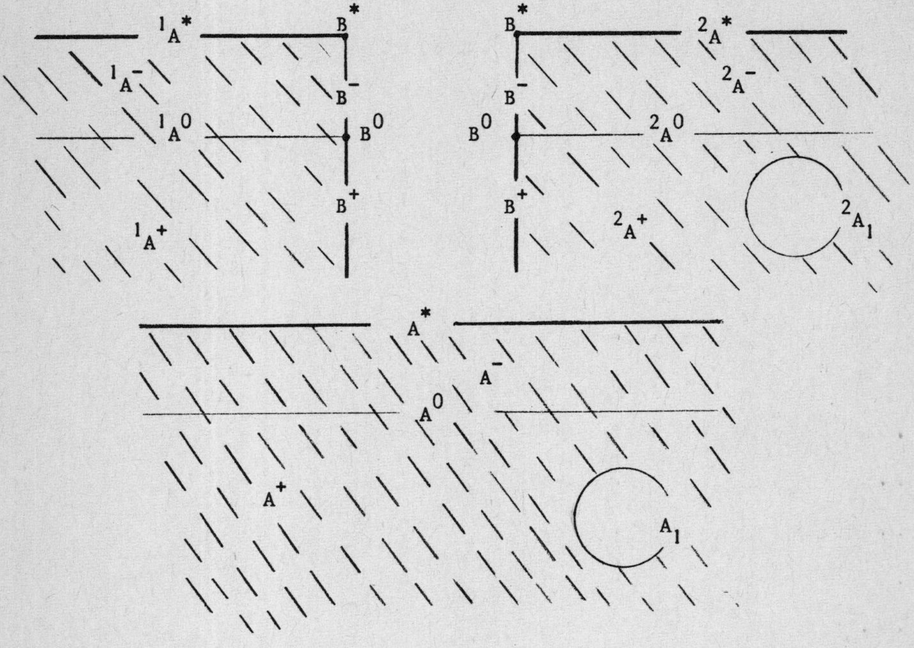

6.1.9. Let $\underline{\underline{M}}$ be an a.s. of depth zero and let $q : \underline{\underline{A}} \dashrightarrow \underline{\underline{M}}$. A decomposition Δ of $\underline{\underline{A}}$ is called q-<u>compatible</u> if

$$q = q \circ \pi_\Delta \quad \text{near } A^-$$

or, equivalently

$$dq \cdot \xi_\Delta = 0 \quad \text{near } A^- .$$

If Δ is q-compatible then $q|A^+$ is a weak morphism from $\underline{\underline{A}}^+$ to $\underline{\underline{M}}$ (if Δ is not q-compatible it may happen that $q|A^+$ is not compatible with the face A^0 of $\underline{\underline{A}}^+$).

6.1.10. Consider a diagram

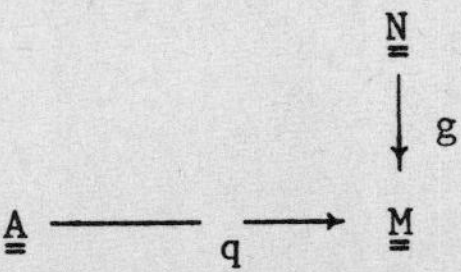

in which $\operatorname{depth}(\underline{\underline{M}}) = 0 = \operatorname{depth}(\underline{\underline{N}})$ and q and g are transverse. Let $\tilde{\underline{\underline{A}}} = \underline{\underline{A}} \times_{\underline{\underline{M}}} \underline{\underline{N}}$ be the fibre product of $\underline{\underline{A}}$ and $\underline{\underline{N}}$ with respect to q and g (see 5.9.3); since $\operatorname{depth}(\underline{\underline{N}}) = 0$, $\tilde{A}$ is an a.s. Let $\Delta = \{\underline{\underline{A}}^-, \underline{\underline{A}}^+, p, \varepsilon, \Phi\}$ be a q-compatible decomposition of $\underline{\underline{A}}$. We shall define a decomposition $\Delta^g = \{\tilde{\underline{\underline{A}}}^-, \tilde{\underline{\underline{A}}}^+, p^g, \varepsilon^g, \Phi^g\}$ of $\tilde{\underline{\underline{A}}}$ as follows. Notice first that $q|A^-: \underline{\underline{A}}^- \dashrightarrow \underline{\underline{M}}$ (resp. $q|A^+: \underline{\underline{A}}^+ \dashrightarrow \underline{\underline{M}}$, $q|A^*: \underline{\underline{A}}^* \dashrightarrow \underline{\underline{M}}$, $q|A^0: \underline{\underline{A}}^0 \dashrightarrow \underline{\underline{M}}$) and g are transverse and thus we can consider $\tilde{\underline{\underline{A}}}^- = \underline{\underline{A}}^- \times_M \underline{\underline{N}}$ (resp. $\tilde{\underline{\underline{A}}}^+ = \underline{\underline{A}}^+ \times_M \underline{\underline{N}}$, $\tilde{\underline{\underline{A}}}^* = \underline{\underline{A}}^* \times_M \underline{\underline{N}}$, $\tilde{\underline{\underline{A}}}^0 = \underline{\underline{A}}^0 \times_M \underline{\underline{N}}$) and identify its underlying topological space with a closed subset of $\tilde{A}$. Then $\tilde{\underline{\underline{A}}}^* = (\tilde{\underline{\underline{A}}})^*$, $\tilde{A}^0 = \tilde{A}^- \cap \tilde{A}^+$ is a face of both $\tilde{\underline{\underline{A}}}^-$ and $\tilde{\underline{\underline{A}}}^+$ and $\tilde{\underline{\underline{A}}}^- | \tilde{A}^0 = \tilde{\underline{\underline{A}}}^+ | \tilde{A}^0$. Define $p^g\colon \tilde{A}^0 \longrightarrow \tilde{A}^*$, $\varepsilon^g\colon \tilde{A}^* \longrightarrow R_+^*$ and $\Phi^g\colon C(p^g) \longrightarrow \tilde{A}^-$ by setting: $p^g(a,x) = (p(a),x)$, $(a,x) \in \tilde{A}^0 = A^0 \times_M N$; $\varepsilon^g(a,x) = \varepsilon(a)$, $(a,x) \in \tilde{A}^* = A^* \times_M N$; $\Phi^g([(a,x),t]) = ([(a,t]),x)$, $[(a,x),t] \in C(p^g)$. A direct verification shows that Δ^g so defined is a decomposition of $\tilde{\underline{\underline{A}}}$. If U_Δ is small enough, we can take $U_{\Delta^g} = (U_\Delta \times N) \cap \tilde{A}$ and then

$$\varphi_{\Delta^g}(a,x) = \varphi_\Delta(a), \quad (a,x) \in U_{\Delta^g};$$

$$\pi_{\Delta^g}(a,x) = (\pi_\Delta(a),x), \quad (a,x) \in U_{\Delta^g};$$

$$\xi_{\Delta^g} = (\xi_\Delta \times 0)|U_{\Delta^g} \setminus \tilde{A}^* .$$

6.2. THEOREM. Let $\underline{\underline{A}}$ and $\underline{\underline{M}}$ be a.s.'s, $0 \leq \operatorname{depth}(\underline{\underline{A}}) < \infty$, $\operatorname{depth}(\underline{\underline{M}}) = 0$ and let $q : \underline{\underline{A}} \dashrightarrow \underline{\underline{M}}$. Suppose that there exists $I \subset I_{\underline{\underline{A}}}$ such that, for any $i \in I$,

(a) $\operatorname{depth}(\underline{\underline{A}}_i) = \operatorname{depth}(\underline{\underline{A}})$;

(b) there is given a $(q|A_i)$-compatible decomposition $\Delta_i = \{\underline{\underline{A}}_i^-, \underline{\underline{A}}_i^+, p_i, \varepsilon_i, \Phi_i\}$ of $\underline{\underline{A}}_i = \underline{\underline{A}}|A_i$;

(c) if $j \in I$, $j \neq i$, $A_i \cap A_j \neq \emptyset$ and $\operatorname{depth}(\underline{\underline{A}}_i|A_i \cap A_j) = \operatorname{depth}(\underline{\underline{A}}_i)$ then $\Delta_i|A_i \cap A_j = \Delta_j|A_i \cap A_j$ (note that $A_i \cap A_j$ is a face of both $\underline{\underline{A}}_i$ and $\underline{\underline{A}}_j$).

Then there exists a q-compatible decomposition Δ of $\underline{\underline{A}}$ such that $\Delta|A_i = \Delta_i$, $i \in I$.

<u>Proof</u>. For any $i \in I$ set

$$V_i^* = A^* \cap U_{A_i} \quad \text{and} \quad V_i = F_{A_i}^{-1}((\pi_{\Delta_i} \times 1_{R_+})^{-1}(F_{A_i}(V_i^*))).$$

Define $\pi_i : V_i \to V_i^*$ and $\varphi_i : V_i \to R_+$ to be the unique maps which make commutative the diagrams

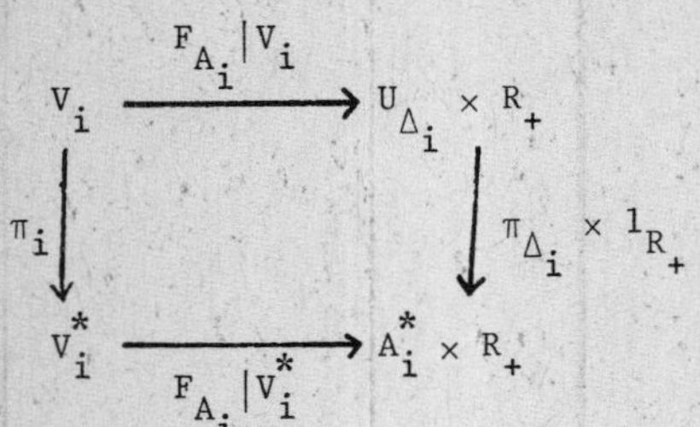

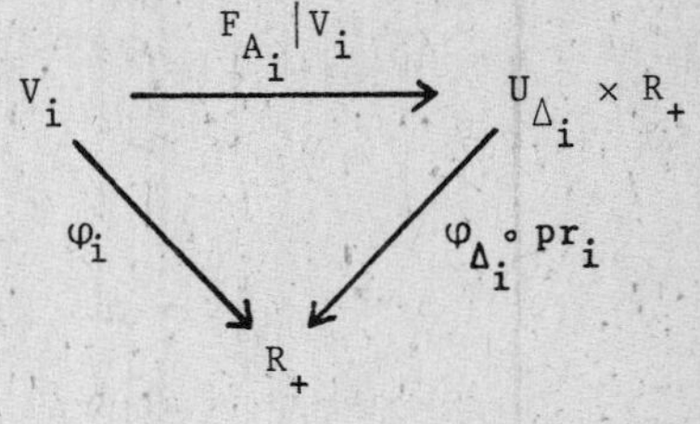

where pr_i is the projection on U_{Δ_i} .

Consider $j \in I$, $j \neq i$. Taking into account (c), we can assume (possibly after shrinking the U_{A_k} 's and the U_{Δ_k} 's, $k \in I$) that

$$\pi_i|V_i \cap V_j = \pi_j|V_i \cap V_j$$

and

$$\varphi_i|V_i \cap V_j = \varphi_j|V_i \cap V_j \,.$$

Next let $\alpha : A^* \to R_+^*$ be such that for any $X \in A^*$ all the control conditions involved hold on $T_X^{\alpha|X}$. By taking α sufficiently small and (if necessary) by shrinking once more the U_{A_k} 's and the U_{Δ_k} 's ($k \in I$) we may assume that for any $i \in I$

$$\pi^{\alpha)}|A^{\alpha)} \cap V_i = \pi_i|A^{\alpha)} \cap V_i$$

$$\rho^{\alpha)}|A^{\alpha)} \cap V_i = (\varphi_i\tilde{\varepsilon}_i)|A^{\alpha)} \cap V_i$$

where $\tilde{\varepsilon}_i = \varepsilon_i \circ \pi_{\Delta_i} \circ (p_{A_i}|V_i) : V_i \to R_+$. Set

$$U = A^{\alpha)} \cup (\bigcup_{i \in I} V_i)$$

and define $\pi : U \to A^*$ and $\rho : U \to R_+$ by setting

$$\pi|A^{\alpha)} = \pi^{\alpha)} , \quad \pi|V_i = \pi_i$$

$$\rho|A^{\alpha)} = \rho^{\alpha)} , \quad \rho|V_i = \varphi_i \tilde{\varepsilon}_i .$$

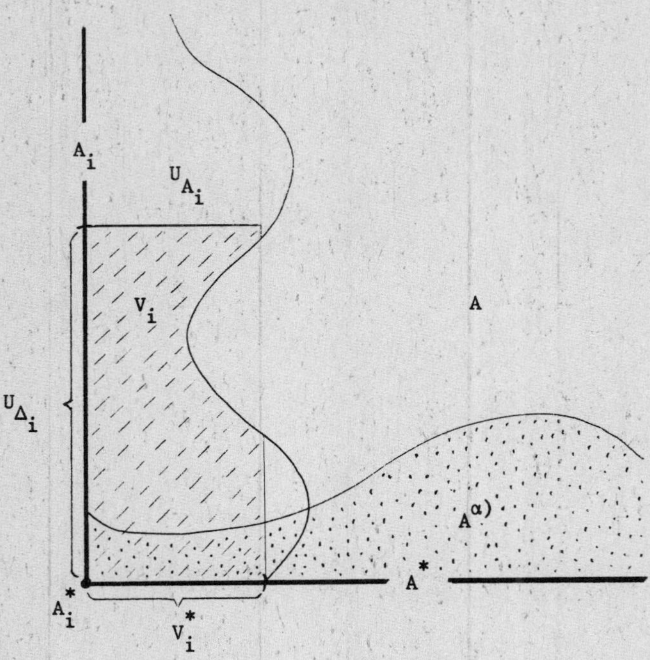

Notice that

$$(A_i^* \times [0,\varepsilon_i]) \cup (A^* \times \{0\}) \subset \{(\pi(a),\rho(a));\ a \in U\} \subset A^* \times R_+$$

and the set in the middle is open in $A^* \times R_+$. Hence there exists $\varepsilon : \underline{\underline{A}}^* \dashrightarrow \underline{\underline{R}}_+^*$ such that

$$\varepsilon|A_i = \varepsilon_i , \quad i \in I$$

and

$$A^* \times [0,\varepsilon] \subset \{(\pi(a),\rho(a));\ a \in U\} .$$

Let $\varphi = \rho/\varepsilon \circ \pi : U \to R_+$. Clearly $\varphi|U_{\Delta_i} = \varphi_{\Delta_i}$ and $\varphi|A^{\alpha)} = \rho^{\alpha)}/\varepsilon \circ \pi^{\alpha)}$. Set $W = \{(\pi(a), \varphi(a)) \in A^* \times R_+^* ; a \in U \smallsetminus A^*\}$; then W is open in $A^* \times R_+^*$ and $(\underline{\underline{A}}^* \times \underline{\underline{R}}_+^*)|W$ is an a.s. of depth zero. Define $f : U \smallsetminus A^* \to W$ by $f(a) = (\pi(a), \varphi(a))$. It is easily checked that f is a proper, submersive weak morphism of $\underline{\underline{A}}|U \smallsetminus A^*$ on $(\underline{\underline{A}}^* \times \underline{\underline{R}}_+^*)|W$; by 5.9.4 f is an a.T.m. Notice that $U_{\Delta_i} \smallsetminus A_i^* = (U \smallsetminus A^*) \cap A_i$ is a face of $\underline{\underline{A}}|U \smallsetminus A^*$ and $df \cdot \xi_{\Delta_i} = 0 \times dt$. By Proposition 5.6, there exists $\xi \in \mathcal{X}_{\underline{\underline{A}}}(U \smallsetminus A^*)$ such that

(6.2.1) $$\xi|U_{\Delta_i} \smallsetminus A_i^* = \xi_{\Delta_i}$$

and

$$df \cdot \xi = 0 \times d/dt \in \mathcal{X}_{\underline{\underline{A}}^* \times \underline{\underline{R}}_+^*}(W) .$$

This last relation is equivalent to

(6.2.2) $$d\varphi \cdot \xi = d/dt$$

and

(6.2.3) $$d\pi \cdot \xi = 0 .$$

From the choice of ε and the definitions of φ and W, we deduce that $A^* \times (0,1] \subset W$. Then $A^- = \varphi^{-1}((-\infty,1])$ is closed in A. Let $A^0 = \varphi^{-1}(1)$ and $A^+ = (A \smallsetminus A^-) \cup A^0$. By Proposition 5.10, we can endow A^- and A^+ with a.s. structures $\underline{\underline{A}}^-$ and $\underline{\underline{A}}^+$ having the properties required in 6.1.1(i) and such that the vector field associated to the face A^0 of $\underline{\underline{A}}^-$ (resp. $\underline{\underline{A}}^+$) is a restriction of ξ.

Let $p = \pi|A^0 : A^0 \to A^*$. Then p is a proper, submersive weak morphism from $\underline{\underline{A}}^0 = \underline{\underline{A}}|A^0$ to $\underline{\underline{A}}^*$ sending strata onto strata. We can therefore consider $\underline{\underline{C}}(p)^{\varepsilon}$.

Let $\lambda_\xi : D_\xi \to U \smallsetminus A^*$ be the flow associated to ξ. Since f is proper it follows that

$$\{(a,t) \in (U \setminus A^*) \times R;\ -\varphi(a) < t \leq 0\} = \{(a,t) \in D_\xi;\ t \leq 0\}.$$

Taking into account (6.2.2), (6.2.3) and (1.1.3) we obtain

$$\lim_{t \nearrow \varphi(a)} \lambda_\xi(a,-t) = \pi(a), \quad a \in U \setminus A^* .$$

Define $\Phi : C(p) \to A^-$ by

$$\Phi([a,t]) = \lambda_\xi(a,t-1), \quad a \in A^0, \quad 0 < t \leq 1 ;$$

$$\Phi([a,0]) = p(a), \quad a \in A^0 ;$$

$$\Phi([a,0]) = a, \quad a \in A^* .$$

Then Φ is an isomorphism of $\underline{\underline{C}}(p)^\varepsilon$ on $\underline{\underline{A}}^-$. It follows that $\Delta = \{\underline{\underline{A}}^-, \underline{\underline{A}}^+, \varepsilon, p, \Phi\}$ is a decomposition of $\underline{\underline{A}}$. From the construction, $\Delta|A_i = \Delta_i$, $i \in I$. Finally, by choosing ε carefully, we may assume that $dq \cdot \xi = 0$ near A^-, which means that Δ is q-compatible. Q.E.D.

<u>Remark</u>. If in the above theorem $I = \emptyset$, then the construction of Δ is much simpler. Indeed, in this case, $U = A^{\alpha)}$, $\pi = \pi^{\alpha)}$, $\varphi = \rho^{\alpha)}/\varepsilon \circ \pi^{\alpha)}$, $\varepsilon : A^* \to R_+^*$ being any controlled map such that $\varepsilon < \alpha$, and all the assertions made during the proof of Theorem 6.2 are obvious.

6.3.1. Recall that $0 \leq \text{depth}(\underline{\underline{A}}) < \infty$. We define inductively the notion of <u>total decomposition</u> of $\underline{\underline{A}}$, denoted $\mathcal{D}$, as follows. If $\text{depth}(\underline{\underline{A}}) = 0$, then $\mathcal{D} = \{\Delta^1\}$, where Δ^1 is a decomposition of $\underline{\underline{A}}$. If $n = \text{depth}(\underline{\underline{A}}) > 0$, then $\mathcal{D} = \{\Delta^1, \Delta^2, \ldots, \Delta^{n+1}\}$ where $\Delta^1 = \{\underline{\underline{A}}^-, \underline{\underline{A}}^+, p, \varepsilon, \Phi\}$ is a decomposition of $\underline{\underline{A}}$ and $\mathcal{D}^+ = \{\Delta^2, \Delta^3, \ldots, \Delta^{n+1}\}$ is a total decomposition of $\underline{\underline{A}}^+$.

6.3.2. Let $\mathcal{D} = \{\Delta^1, \ldots, \Delta^{n+1}\}$ be a total decomposition of $\underline{\underline{A}}$ and A_i be a face of $\underline{\underline{A}}$ such that $\text{depth}(\underline{\underline{A}}_i) = n-k \geq 0$. Then $\mathcal{D}$ induces a total decomposition of $\underline{\underline{A}}_i$, denoted $\mathcal{D}|A_i$, as follows. Let $\underline{\underline{A}}^j$ be the a.s. of which Δ^j is a decomposition; then $\text{depth}(\underline{\underline{A}}^j) = n-j+1$ (this follows easily by induction on j). Since $\text{depth}(\underline{\underline{A}}_i) = n-k$ it follows that A_i is a face of $\underline{\underline{A}}^{k+1}$ and we can consider $\Delta^{k+1}|A_i$. Moreover $A_i \cap A^j$ is a non empty face of $\underline{\underline{A}}^j$ for all $j > k+1$ and $\operatorname{depth}(\underline{\underline{A}}^j | A_i \cap A^j) = \operatorname{depth}(\underline{\underline{A}}^j)$. Define $\mathcal{D}|A_i$ by

$$\mathcal{D}|A_i = \{\Delta^{k+1}|A_i, \Delta^{k+2}|A_i \cap A^{k+2}, \ldots, \Delta^{n+1}|A_i \cap A^{n+1} .$$

An inductive verification shows that $\mathcal{D}|A_i$ is indeed a total decomposition of $\underline{\underline{A}}_i$.

6.3.3. Let $f\colon \underline{\underline{A}} \dashrightarrow \underline{\underline{M}}$, $\operatorname{depth}(\underline{\underline{M}}) = 0$. A total decomposition $\mathcal{D} = \{\Delta^1, \Delta^2, \ldots, \Delta^{n+1}\}$ of $\underline{\underline{A}}$ is called f-<u>compatible</u> if Δ^1 is an f-compatible decomposition of $\underline{\underline{A}}$ and, inductively if $n > 0$, the total decomposition $\mathcal{D}^+ = \{\Delta^2, \ldots, \Delta^{n+1}\}$ of $\underline{\underline{A}}^+$ is $f|A^+$-compatible.

6.3.4. A total decomposition $\mathcal{D} = \{\Delta^1, \Delta^2, \ldots, \Delta^{n+1}\}$ of $\underline{\underline{A}}$ is called <u>regular</u> if either $n = 0$ or $n > 0$ and then (inductively) the total decomposition $\mathcal{D}^+ = \{\Delta^2, \ldots, \Delta^{n+1}\}$ of $\underline{\underline{A}}^+$ is regular and $\mathcal{D}^+|A^0$ is a p-compatible total decomposition of $\underline{\underline{A}}^0$ (recall that $A^0 = A^- \cap A^+$ is a face of $\underline{\underline{A}}^+$).

6.3.5. Let

$$\begin{array}{ccc} & & \underline{\underline{N}} \\ & & \downarrow g \\ \underline{\underline{A}} & \xrightarrow{q} & \underline{\underline{M}} \end{array}$$

be as in 6.1.10 and let $\mathcal{D} = \{\Delta^1, \Delta^2, \ldots, \Delta^{n+1}\}$ be a q-compatible total decomposition of $\underline{\underline{A}}$. We define a total decomposition $\mathcal{D}^g = \{\tilde{\Delta}^1, \tilde{\Delta}^2, \ldots, \tilde{\Delta}^{n+1}\}$ of $\tilde{\underline{\underline{A}}} = \underline{\underline{A}} \times_M \underline{\underline{N}}$ as follows. Set $\tilde{\Delta}^1 = \Delta^g$ (see 6.1.10) and, inductively if $n > 0$, $\{\tilde{\Delta}^2, \ldots, \tilde{\Delta}^{n+1}\} = (\mathcal{D}^+)^g$.

The following lemma is an immediate consequence of the definitions introduced above.

6.4. LEMMA. Let $f\colon \underline{\underline{A}} \dashrightarrow \underline{\underline{M}}$ with $\operatorname{depth}(\underline{\underline{M}}) = 0$, let $\mathcal{D}$ be a total decomposition of $\underline{\underline{A}}$ and let A_i be a non empty face of $\underline{\underline{A}}$. If $\mathcal{D}$ is regular (resp. f-compatible) then $\mathcal{D}|A_i$ is regular (resp. f_i-compatible) (here $f_i\colon A_i \to M_i$ is the restriction of f).

6.5. THEOREM. Let $f\colon \underline{\underline{A}} \dashrightarrow \underline{\underline{M}}$ with depth($\underline{\underline{M}}$) = 0. Let $I \subset I_{\underline{\underline{A}}}$ and for any $i \in I$ let $\mathcal{D}_i$ be a regular and f_i-compatible total decomposition of $\underline{\underline{A}}_i$. Assume that $\mathcal{D}_i|A_i \cap A_j = \mathcal{D}_j|A_i \cap A_j$ if $A_i \cap A_j \neq \emptyset$ $(i,j \in I)$. Then there exists a regular and f-compatible total decomposition $\mathcal{D}$ of $\underline{\underline{A}}$ such that $\mathcal{D}|A_i = \mathcal{D}_i$ for any $i \in I$.

Proof. The proof is simple and we shall only sketch it. If depth($\underline{\underline{A}}$) = 0, there is nothing to prove. Assume that depth($\underline{\underline{A}}$) > 0. Construct first an f-compatible decomposition $\Delta^1 = \{\underline{\underline{A}}^-, \underline{\underline{A}}^+, p, \varepsilon, \Phi\}$ of $\underline{\underline{A}}$ with the required properties. Next construct by induction on depth($\underline{\underline{A}}$) a regular and p-compatible total decomposition $\mathcal{D}^0$ of $\underline{\underline{A}}^0$ such that $\mathcal{D}^0|A^0 \cap A_i = \mathcal{D}_i|A^0 \cap A_i$ if $A^0 \cap A_i \neq \emptyset$, $i \in I$. Notice that $\mathcal{D}^0$ is also f-compatible since Δ^1 is f-compatible and thus $f|A^0 = (f|A^*)\circ p$. Next construct, again by induction, a regular and $(f|A^+)$-compatible total decomposition $\mathcal{D}^+ = \{\Delta^2, \ldots, \Delta^{n+1}\}$ of $\underline{\underline{A}}^+$ such that $\mathcal{D}^+|A^0 = \mathcal{D}^0$ and $\mathcal{D}^+|A_i \cap A^+ = \mathcal{D}_i|A_i \cap A^+$ when this makes sense. Finally set $\mathcal{D} = \{\Delta^1, \Delta^2, \ldots, \Delta^{n+1}\}$. Q.E.D.

6.6.1. Let $\mathcal{D} = \{\Delta^1, \Delta^2, \ldots, \Delta^{n+1}\}$ be a total decomposition of $\underline{\underline{A}}$ with $\Delta^1 = \{\underline{\underline{A}}^-, \underline{\underline{A}}^+, p, \varepsilon, \Phi\}$. If $n = 0$ set $c(\underline{\underline{A}},\mathcal{D}) = A$. If $n > 0$ define inductively $c(\underline{\underline{A}},\mathcal{D}) = A^* \sqcup c(\underline{\underline{A}}^+,\mathcal{D}^+)$, where $\mathcal{D}^+ = \{\Delta^2, \ldots, \Delta^{n+1}\}$. It is obvious from the definition that $c(\underline{\underline{A}},\mathcal{D})$ is a manifold with faces; it is called the $\mathcal{D}$-core of $\underline{\underline{A}}$.

6.6.2. Given two total decompositions $\mathcal{D}$ and $\mathcal{D}'$ of $\underline{\underline{A}}$, it is easily seen that $c(\underline{\underline{A}},\mathcal{D})$ and $c(\underline{\underline{A}},\mathcal{D}')$ are diffeomorphic (we shall not use this assertion).

6.6.3. Let $\mathcal{D}$ be a total decomposition of $\underline{\underline{A}}$ and A_i be a face of $\underline{\underline{A}}$. Then an easy inductive argument shows that $A_i \cap c(\underline{\underline{A}},\mathcal{D}) = c(\underline{\underline{A}}_i,\mathcal{D}|A_i)$ and that $c(\underline{\underline{A}}_i,\mathcal{D}|A_i)$ is a face of $c(\underline{\underline{A}},\mathcal{D})$.

6.7. From now on $\underline{\underline{M}}$ will be an a.s. of depth zero and $p: \underline{\underline{A}} \dashrightarrow \underline{\underline{M}}$ will be a proper submersive weak morphism sending strata onto strata, all the faces of $\underline{\underline{A}}$ being p-vertical (recall that $0 \leq \operatorname{depth}(\underline{\underline{A}}) < \infty$). Some of these conditions are not necessary in all what follows, but since they are necessary for the main constructions we prefer to consider them from the beginning.

6.7.1. A commutative square

$$(S) \qquad \begin{array}{ccc} \underline{\underline{B}} & \xrightarrow{q} & \underline{\underline{N}} \\ f \downarrow & & \downarrow g \\ \underline{\underline{A}} & \xrightarrow[p]{} & \underline{\underline{M}} \end{array}$$

is called __regular__ if

(i) $\underline{\underline{N}}$ is an a.s. of depth zero;

(ii) for any stratum Y of $\underline{\underline{B}}$ let X (resp. Z) be the stratum of $\underline{\underline{A}}$ (resp. $\underline{\underline{N}}$) containing $f(Y)$ (resp. $q(Y)$); then the diagram

$$\begin{array}{ccc} Y & \xrightarrow{q|Y} & Z \\ f|Y \downarrow & & \downarrow g|Z \\ X & \xrightarrow[p|X]{} & p(X) \end{array}$$

is regular (see 3.1.).

This last condition is easily seen to be equivalent to

(ii') let $\tilde{\underline{\underline{A}}} = \underline{\underline{A}} \times_M \underline{\underline{N}}$ be the fibre product of $\underline{\underline{A}}$ and $\underline{\underline{N}}$ with respect to p and g and define $\tilde{f}: \underline{\underline{B}} \dashrightarrow \tilde{\underline{\underline{A}}}$ by $\tilde{f}(b) = (f(b), q(b))$; then $\tilde{f}$ is submersive.

__Remark.__ The notation being as above, $\tilde{f}$ is an a.T.m. and the square

$$(\tilde{S}) \qquad \begin{array}{ccc} \underline{\underline{B}} & \xrightarrow{q} & \underline{\underline{N}} \\ \tilde{f} \downarrow & & \downarrow 1_N \\ \tilde{\underline{\underline{A}}} & \xrightarrow[p]{} & \underline{\underline{N}} \end{array} \qquad \tilde{p}(a,x) = x$$

is regular.

If S is regular and in addition

(iii) q is proper, submersive and sends strata onto strata;

(iv) $\underline{\underline{B}}$ has at most one q-horizontal face B^h (which is necessarily f-horizontal),

then S is called <u>basic</u>.

If S is basic and all the faces of $\underline{\underline{B}}$ are q-vertical, then S is called <u>admissible of depth zero</u>.

Let $n > 0$ and assume inductively that we have defined the notion of admissible square of depth $n-1$. A commutative square

(S) 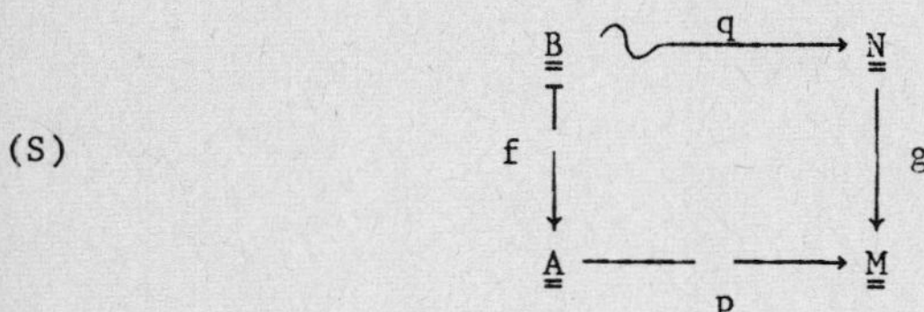

is called <u>admissible of depth</u> n if

(1) $\underline{\underline{N}}$ is an a.s. of depth n;

(2) there is given a g-compatible decomposition $\Delta_S = \{\underline{\underline{N}}^-, N^+, p, \varepsilon, \Phi\}$ of $\underline{\underline{N}}$;

(3) there exist closed subsets B^-, $B^\times$, and B^+ of B, endowed with w.a.s. structures $\underline{\underline{B}}^-$, $\underline{\underline{B}}^\times$ and $\underline{\underline{B}}^+$ respectively such that $f^- = f|B^- : B^- \to A$ (resp. $f^\times = f|B^\times : B^\times \to A$, resp. $f^+ = f|B^+ : B^+ \to A$) is an a. T. m. from $\underline{\underline{B}}^-$ (resp. $\underline{\underline{B}}^\times$, resp. $\underline{\underline{B}}^+$) to $\underline{\underline{A}}$; $B^0 = B^+ \cap B^\times$ is an f^+-horizontal face of $\underline{\underline{B}}^+$ and an $f^\times$-horizontal face of $\underline{\underline{B}}^\times$ and $\underline{\underline{B}}^+|B^0 = \underline{\underline{B}}^\times|B^0$; $B^h = B^\times \cap B^-$ is an $f^\times$-horizontal face of $\underline{\underline{B}}^\times$ and an f^--horizontal face of $\underline{\underline{B}}^-$ and $\underline{\underline{B}}^\times|B^h = \underline{\underline{B}}^-|B^h$; $B^+ \cap B^- = \emptyset$ and one can consider $\underline{\underline{B}}^- \cup_{B^h} \underline{\underline{B}}^\times \cup_{B^0} \underline{\underline{B}}^+$; the inclusions of B^-, $B^\times$ and B^+ in B determine an isomorphism from $f^- \cup_{B^h} f^\times \cup_{B^0} f^+ : \underline{\underline{B}}^- \cup_{B^h} \underline{\underline{B}}^\times \cup_{B^0} \underline{\underline{B}}^+ \vdash\!\to \underline{\underline{A}}$ to $f: \underline{\underline{B}} \vdash\!\to \underline{\underline{A}}$ over 1_A ;

(4) $B^+ = q^{-1}(N^+)$, $q^+ = q|B^+ : B^+ \to N^+$ is a quasimorphism, $q^{-1}(N^0) = B^0$ and the square

$$(S^+) \qquad \begin{array}{ccc} \underline{\underline{B}}^+ & \overset{q^+}{\rightsquigarrow} & \underline{\underline{N}}^+ \\ f^+ \downarrow & & \downarrow g^+ = g|N^+ \\ \underline{\underline{A}} & \underset{p}{\longrightarrow} & \underline{\underline{M}} \end{array}$$

is admissible of depth $n - 1$;

(5) $B^- = q^{-1}(N^*)$, $q^- = q|B^- : B^- \to N^*$ is a weak morphism and the square

$$(S^-) \qquad \begin{array}{ccc} \underline{\underline{B}}^- & \overset{q^-}{\longrightarrow} & \underline{\underline{N}}^* \\ f^- \downarrow & & \downarrow g^* = g|N^* \\ \underline{\underline{A}} & \underset{p}{\longrightarrow} & \underline{\underline{M}} \end{array}$$

is basic, the q^--horizontal face of $\underline{\underline{B}}^-$ being B^h ;

(6) let $\underline{\underline{B}}^h = \underline{\underline{B}}|B^h = \underline{\underline{B}}^-|B^h = \underline{\underline{B}}^\times|B^h$ and $f : \underline{\underline{B}}^h \times [0,1] \vdash\!\to \underline{\underline{A}}$ be the a.T.m. given by $\tilde{f}(b,t) = f(b)$; then there exists a weak isomorphism $\theta_S : \underline{\underline{B}}^h \times [0,1] \dashrightarrow \underline{\underline{B}}^\times$ such that $\theta_S(b,0) = b$, $\theta_S(B^h \times \{1\}) = B^0$, $q(\theta_S(b,t)) = \Phi([q(b),t])$ (thus $q(B^\times) \subset N^-$) and $(\theta_S, 1_A)$ is an isomorphism from $\tilde{f}$ to $f^\times$.

(7) Let $\mathcal{D}_S$ be the g-compatible total decomposition of $\underline{\underline{N}}$ defined inductively as follows: if $n = 0$ $\mathcal{D}_S$ is the obvious total decomposition of the a.s. of depth zero $\underline{\underline{N}}$; if $n > 0$ set $\mathcal{D}_S = \{\Delta_S\} \cup \mathcal{D}_{S^+}$. Then $\mathcal{D}_S$ is regular.

Now it is easy to check by induction that

(8) q is proper;

(9) all the faces of $\underline{\underline{B}}$ are q-vertical;

(10) for any $Y \in \mathcal{B}$, $q(Y)$ is closed in N ;

(11) for any strata $X \in A$, $Y \in B$ and $Z \in N$ with $f(Y) \subset X$ and $Z \subset q(Y)$ there exists α such that the diagram

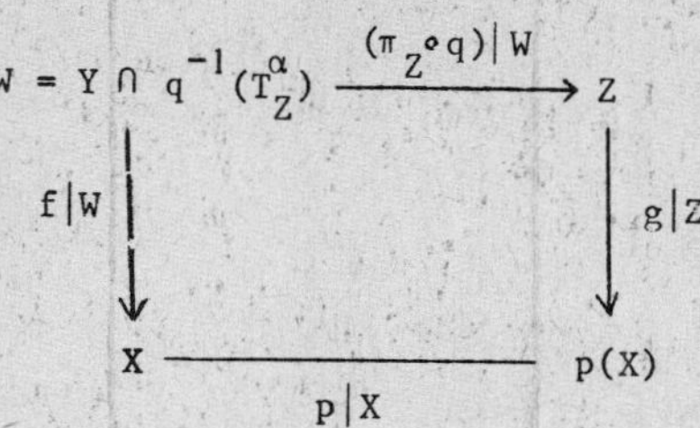

is regular.

(12) let $\tilde{M}$ be an a.s. of depth zero and let $\tilde{g} : \underline{\underline{N}} \dashrightarrow \tilde{\underline{\underline{M}}}$ and $\tilde{\tilde{g}} : \tilde{\underline{\underline{M}}} \dashrightarrow \underline{\underline{M}}$ be submersive weak morphisms such that $g = \tilde{\tilde{g}} \circ \tilde{g}$. Let $\tilde{\underline{\underline{A}}} = \underline{\underline{A}} \times_M \tilde{\underline{\underline{M}}}$ (with respect to p and $\tilde{\tilde{g}}$) and define $\tilde{f} : B \to \tilde{A}$ and $\tilde{p} : \tilde{A} \to \tilde{M}$ by $\tilde{f}(b) = (f(b), \tilde{g}(q(b)))$ and $\tilde{p}(a,x) = x$. Then $\tilde{f}$ is an a.T.m. from $\underline{\underline{B}}$ to $\tilde{\underline{\underline{A}}}$, $\tilde{p}$ is a proper, submersive, weak morphism sending strata onto strata, all the faces of $\tilde{\underline{\underline{A}}}$ are $\tilde{p}$-vertical and, if $\mathcal{D}_S$ is $\tilde{g}$-compatible, S induces a structure of admissible square of depth n on

$$(\tilde{S}) \qquad \begin{array}{ccc} \underline{\underline{B}} & \overset{q}{\rightsquigarrow} & \underline{\underline{N}} \\ {\scriptstyle \tilde{f}}\downarrow & & \downarrow{\scriptstyle \tilde{g}} \\ \tilde{\underline{\underline{A}}} & \underset{\tilde{p}}{\dashrightarrow} & \tilde{\underline{\underline{M}}} \end{array}$$

as follows: $\Delta_{\tilde{S}} = \Delta_S$, $\theta_{\tilde{S}} = \theta_S$ and the squares $\tilde{S}^+$ and $\tilde{S}^-$ are respectively

$$\begin{array}{ccc} \underline{\underline{B}}^+ & \overset{q^+}{\rightsquigarrow} & \underline{\underline{N}}^+ \\ {\scriptstyle \tilde{f}|B^+}\downarrow & & \downarrow{\scriptstyle \tilde{g}|N^+} \\ \tilde{\underline{\underline{A}}} & \underset{\tilde{p}}{\dashrightarrow} & \tilde{\underline{\underline{M}}} \end{array}$$

and

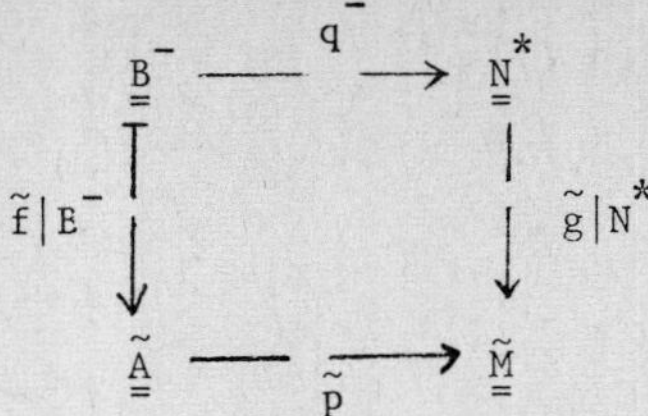

Later it will be useful to consider the squares

$$(S^{\times}) \qquad \begin{array}{ccc} \underline{\underline{B}}^{\times} & \xrightarrow{\;q^{\times}\;} & \underline{\underline{N}}^{-} \\ {\scriptstyle f^{\times}}\downarrow & & \downarrow{\scriptstyle g^{-} = g|N^{-}} \\ \underline{\underline{A}} & \xrightarrow[\;p\;]{} & \underline{\underline{M}} \end{array}$$

and

$$(S_e) \qquad \begin{array}{ccc} \underline{\underline{B}}_e = \underline{\underline{B}}^{-} \cup_{B^h} \underline{\underline{B}}^{\times} & \xrightarrow{\;q_e\;} & \underline{\underline{N}}^{-} \\ {\scriptstyle f_e}\downarrow & & \downarrow{\scriptstyle g^{-}} \\ \underline{\underline{A}} & \xrightarrow[\;p\;]{} & \underline{\underline{M}} \end{array}$$

whose maps are the restrictions of the corresponding maps in S.

From now on the notation introduced in this section will be used without any other mention.

6.7.2. Consider the commutative diagram

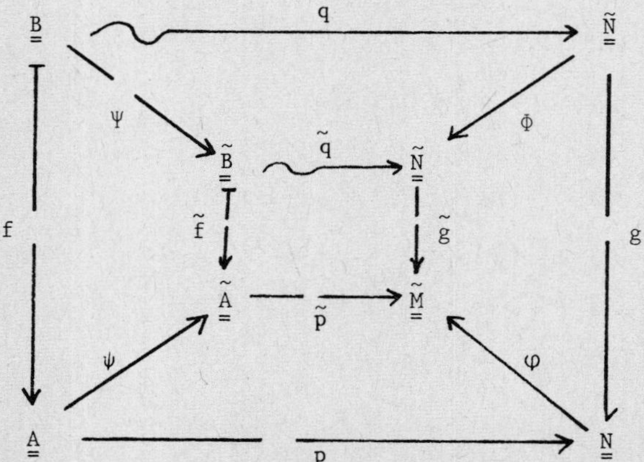

and let S (resp. $\tilde{S}$) denote the exterior (resp. interior) square. The quadruple $(\Psi, \Phi, \psi, \varphi)$ is called an <u>isomorphism</u> from S to $\tilde{S}$ if (Ψ, ψ) is an isomorphism from f to $\tilde{f}$ and Φ and φ are isomorphisms.

Assume now that S and $\tilde{S}$ are admissible squares of depth n. If $n = 0$, an isomorphism from S to $\tilde{S}$ is also called an admissible isomorphism. Assume we have defined inductively the notion of admissible isomorphism between admissible squares of depth $n-1$. An isomorphism $(\Psi, \Phi, \psi, \varphi)$ from S to $\tilde{S}$ is called <u>admissible</u> if

(1) Φ is compatible with Δ and $\tilde{\Delta}$;

(2) $\Psi(B^+) = \tilde{B}^+$, $\Psi(B^\times) = \tilde{B}^\times$ and $\Psi(B^-) = \tilde{B}^-$; let $\Psi^+ : B^+ \to \tilde{B}^+$, $\Psi^\times : B^\times \to \tilde{B}^\times$, $\Psi^- : B^- \to \tilde{B}^-$, $\Psi^h : B^h \to \tilde{B}^h$, $\Phi^+ : N^+ \to \tilde{N}^+$, $\Phi^- : N^- \to \tilde{N}^-$ and $\Phi^* : N^* \to \tilde{N}^*$ be the restrictions of Ψ and Φ respectively;

(3) $(\Psi^+, \Phi^+, \psi, \varphi)$ is an admissible isomorphism from S^+ to $\tilde{S}^+$ (this makes sense by induction);

(4) $(\Psi^-, \Phi^*, \psi, \varphi)$ is an isomorphism from S^- to $\tilde{S}^-$;

(5) $\Psi^\times \circ \theta_S = \theta_{\tilde{S}} \circ (\Psi^h \times 1_{[0,1]})$.

Two admissible squares of depth n, S and $\tilde{S}$, are called <u>admissibly equal</u>, denoted $S \equiv \tilde{S}$, if $B = \tilde{B}$, $A = \tilde{A}$, $N = \tilde{N}$, $M = \tilde{M}$ and $(1_B, 1_N, 1_A, 1_M)$ is an admissible isomorphism from S to $\tilde{S}$. If $B = \tilde{B}$, $A = \tilde{A}$, $N = \tilde{N}$, $M = \tilde{M}$ and $(1_B, 1_N, 1_A, 1_M)$ is only an isomorphism from S to $\tilde{S}$ (i.e., $f = \tilde{f}$, $q = \tilde{q}$, $g = \tilde{g}$ and $p = \tilde{p}$) then S and $\tilde{S}$ are called <u>equal</u>, denoted $S = \tilde{S}$.

Consider again the diagram from the beginning of this subsection and assume that S is admissible of depth n, $\tilde{S}$ is arbitrary and $(\Psi, \Phi, \psi, \varphi)$ is an isomorphism from S to $\tilde{S}$. Then $\tilde{S}$ has a unique structure of admissible square of depth n such that $(\Psi, \Phi, \psi, \varphi)$ is an admissible isomorphism (the construction is obvious and left to the reader).

6.7.3. Let

(S) 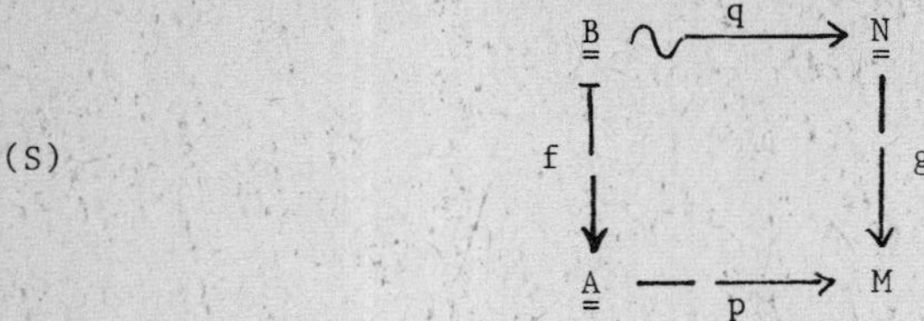

be an admissible square of depth n and B_i be a face of $\underline{\underline{B}}$. Let N_i be the corresponding face of $\underline{\underline{N}}$. Notice that B_i is f-vertical (horizontal) if and only if N_i is g-vertical (horizontal). Consider the square

$$(S|B_i) \qquad \begin{array}{ccc} \underline{\underline{B}}_i & \overset{q_i}{\rightsquigarrow} & \underline{\underline{N}}_i \\ {\scriptstyle f_i}\downarrow & & \downarrow{\scriptstyle g_i} \\ \underline{\underline{A}}_i & \xrightarrow[p_i]{} & \underline{\underline{M}}_i \end{array}$$

whose maps are the restrictions of the corresponding maps in S (if B_i is f-horizontal, then $\underline{\underline{A}}_i = \underline{\underline{A}}$ and $\underline{\underline{M}}_i = \underline{\underline{M}}$; otherwise A_i and M_i are faces of $\underline{\underline{A}}$ and $\underline{\underline{M}}$ respectively and $B_i = f^{-1}(A_i)$, $N_i = g^{-1}(M_i)$, $A_i = p^{-1}(M_i)$). Let $n_i = \text{depth}(\underline{\underline{N}}_i)$. We shall endow $S|B_i$ (by induction on n) with a structure of admissible square of depth n_i as follows. If $n = 0$, then $n_i = 0$ and $S|B_i$ is obviously admissible of depth zero. Suppose $n > 0$.

<u>Case I</u>: $n_i < n$. Then N_i is a face of $\underline{\underline{N}}^+$, B_i is a face of $\underline{\underline{B}}^+$ and $S|B_i = S^+|B_i$. By induction $S^+|B_i$ can be endowed with the required structure and therefore $S|B_i$ too.

<u>Case II</u>: $n_i = n$. Take $\Delta_{S|B_i} = \Delta_S|N_i$. Notice that $B_i^+ = B_i \cap B^+$, $B_i^\times = B_i \cap B^\times$ and $B_i^- = B_i \cap B^-$ are faces of $\underline{\underline{B}}^+$, $B^\times$ and $\underline{\underline{B}}^-$ respectively. Take $\underline{\underline{B}}_i^+ = \underline{\underline{B}}^+|B_i^+$, $\underline{\underline{B}}_i^\times = \underline{\underline{B}}^\times|B_i^\times$ and $\underline{\underline{B}}^-_i = \underline{\underline{B}}^-|B_i^-$. Since

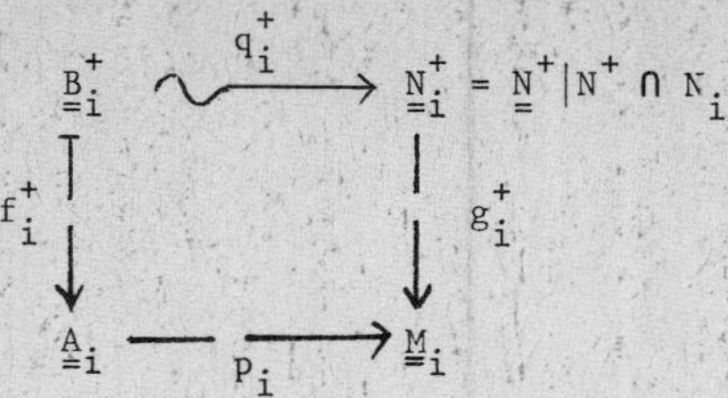

is just $S^+|B_i^+$, it is admissible of depth $n-1$ (by induction). Since $B_i^h = B_i^- \cap B_i^\times = B_i \cap B^h$, we can take $\theta_{S|B_i} : B_i^h \times [0,1] \to B_i^\times$ to be the restriction of θ_S. A direct verification shows that all the conditions involved in the definition of an admissible square are satisfied.

6.7.4. Consider now two admissible squares

$$(S') \qquad \begin{array}{ccc} {}'\underline{\underline{B}} & \xrightarrow{q'} & {}'\underline{\underline{N}} \\ {\scriptstyle f'}\downarrow & & \downarrow{\scriptstyle g'} \\ \underline{\underline{A}} & \xrightarrow[p]{} & \underline{\underline{M}} \end{array}$$

and

$$(S'') \qquad \begin{array}{ccc} {}''\underline{\underline{B}} & \xrightarrow{q''} & {}''\underline{\underline{N}} \\ {\scriptstyle f''}\downarrow & & \downarrow{\scriptstyle g''} \\ \underline{\underline{A}} & \xrightarrow[p]{} & \underline{\underline{M}} \end{array}$$

of depth n' and n'' respectively. Let ${}'B_1$ be an f'-horizontal face of ${}'\underline{\underline{B}}$ and ${}''B_1$ be an f''-horizontal face of ${}''\underline{\underline{B}}$. Suppose that the admissible squares

$$(S'|{}'B_1) \qquad \begin{array}{ccc} {}'\underline{\underline{B}}_1 & \xrightarrow{q_1'} & {}'\underline{\underline{N}}_1 \\ {\scriptstyle f_1'}\downarrow & & \downarrow{\scriptstyle g_1'} \\ \underline{\underline{A}} & \xrightarrow[p]{} & \underline{\underline{M}} \end{array}$$

and

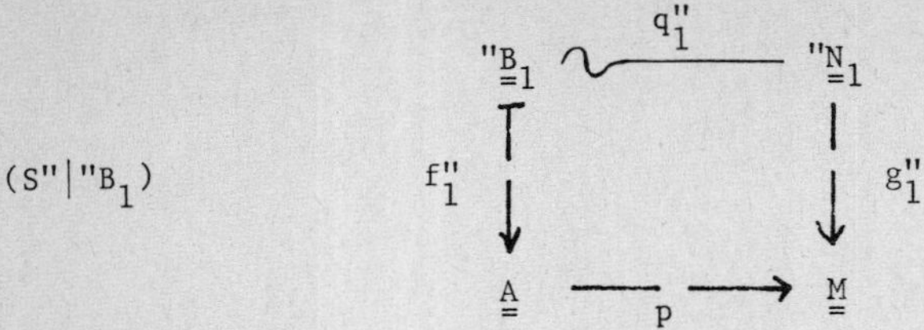

are admissibly equal and that we can construct $'\underline{\underline{B}} \cup_{'B_1} {''\underline{\underline{B}}}$. Let

$\underline{\underline{B}} = {'\underline{\underline{B}}} \cup_{'B_1} {''\underline{\underline{B}}}$, $\underline{\underline{N}} = {'\underline{\underline{N}}} \cup_{'N_1} {''\underline{\underline{N}}}$, $f = f' \cup_{'B_1} f'' : \underline{\underline{B}} \vdash\!\!\rightarrow \underline{\underline{A}}$, $q = q' \cup_{'B_1} q'' : \underline{\underline{B}} \rightsquigarrow \underline{\underline{N}}$ and $g = g' \cup_{'N_1} g'' : \underline{\underline{N}} \dashrightarrow \underline{\underline{M}}$. Consider the square $S = S' \cup_{'B_1} S''$ given by

$$(S) \qquad \begin{array}{ccc} \underline{\underline{B}} & \overset{q}{\rightsquigarrow} & \underline{\underline{N}} \\ f \downarrow & & \downarrow g \\ \underline{\underline{A}} & \xrightarrow[p]{} & \underline{\underline{M}} \end{array}$$

Let $n = \max\{n',n''\}$ and $m = n' + n''$. Using induction on m we shall introduce a structure of admissible square of depth n on S as follows. If $m = 0$, it is clear that S is admissible of depth zero. Assume next that $m > 0$.

<u>Case 1</u>: $n = n' > n''$. Then depth $('\underline{\underline{N}}_1)$ = depth $(''\underline{\underline{N}}_1) \leq n''$ and thus $'N_1 \cap {'N^-} = \emptyset$. Hence $'B_1 \cap {'B^-} = \emptyset = {'B_1} \cap {'B^\times}$ and therefore $'B_1$ is a face of $'\underline{\underline{B}}^+$. Thus

$$'S^+|'B_1 \equiv {'S}|'B_1 \equiv {''S}|''B_1$$

and, by induction, the square $'S^+ \cup_{'B_1} {''S}$ is admissible of depth $n-1$. Take $\Delta_S = \Delta_{'S} \cup_{'N} \Delta_{''S}$, $\underline{\underline{B}}^- = {'\underline{\underline{B}}^-}$, $\underline{\underline{B}}^\times = {'\underline{\underline{B}}^\times}$, $\underline{\underline{B}}^+ = {'\underline{\underline{B}}^+} \cup_{'B_1} {''\underline{\underline{B}}}$ and $S^+ \equiv {'S^+} \cup_{'B_1} {''S}$. Thus S^+ is admissible of depth $n-1$. The other conditions in the definition of an admissible square are easily checked.

<u>Case 2</u>: $n = n'' > n'$. This case is similar to Case 1.

<u>Case 3</u>: $n = n' = n'' >$ depth$('\underline{\underline{N}}_1)$ = depth$(''\underline{\underline{N}}_1)$. Then $'N \cap {'N^-} = \emptyset =$

$= "N_1 \cap "N^-$, hence $'B_1 \cap 'B^- = 'B_1 \cap 'B^\times = \emptyset = "B_1 \cap "B^\times = "B_1 \cap "B^-$ and therefore $'B_1$ is a face of $'\underline{\underline{B}}^+$ and $"B_1$ $(= 'B_1)$ is a face of $"\underline{\underline{B}}^+$. Thus

$$'S^+|'B_1 \equiv 'S|'B_1 \equiv "S|"B_1 \equiv "S^+|"B_1$$

and, by induction, the square $'S^+ \cup_{'B_1} "S^+$ is admissible of depth n-1.

Take $\Delta_S = \Delta_{'S} \cup_{'N_1} \Delta_{"S}$, $\underline{\underline{B}}^- = '\underline{\underline{B}}^- \sqcup "\underline{\underline{B}}^-$, $\underline{\underline{B}}^\times = '\underline{\underline{B}}^\times \sqcup "\underline{\underline{B}}^\times$, $\underline{\underline{B}}^+ = '\underline{\underline{B}}^+ \cup_{'B_1} "\underline{\underline{B}}^+$ and $S^+ \equiv 'S^+ \cup_{'B_1} "S^+$. Thus S^+ is admissible of depth n-1. The other conditions in the definition of an admissible square are easily checked.

<u>Case 4</u>: $n = n' = n'' = \text{depth}('\underline{\underline{N}}_1) = \text{depth}("\underline{\underline{N}}_1)$. In this case $'B_1 \cap 'B^- = 'f^{-1}('N_1 \cap 'N^*) = "f^{-1}("N_1 \cap "N^*) = "B_1 \cap "B^-$. Denote this set by $'B_1^-$. Similarly, let $'B_1^\times = 'B_1 \cap 'B^\times = "B_1 \cap "B^\times$ and $'B_1^+ = 'B_1 \cap 'B^+ = "B_1 \cap "B^+$. Then $'B_1^-$ (resp. $'B_1^\times$, $'B_1^+$) is a face of both $'\underline{\underline{B}}^-$ and $"\underline{\underline{B}}^-$ (resp. $'\underline{\underline{B}}^\times$ and $"\underline{\underline{B}}^\times$, $'\underline{\underline{B}}^+$ and $"\underline{\underline{B}}^+$) and we can construct $\underline{\underline{B}}^- = '\underline{\underline{B}}^- \cup_{'B_1^-} "\underline{\underline{B}}^-$, $\underline{\underline{B}}^\times = '\underline{\underline{B}}^\times \cup_{'B_1^\times} "\underline{\underline{B}}^\times$ and $\underline{\underline{B}}^+ = '\underline{\underline{B}}^+ \cup_{'\underline{\underline{B}}^+} "\underline{\underline{B}}^+$. Notice that

$$'S^+|'B_1^+ \equiv "S^+|'B_1^+$$

and thus, by induction $'S^+ \cup_{'B_1^+} "S^+$ is an admissible square of depth n-1. Take $\Delta_S = \Delta_{'S} \cup_{'N_1} \Delta_{"S}$ and $S^+ \equiv 'S^+ \cup_{'B_1^+} "S^+$. Then S^+ is admissible of depth n-1. The other conditions in the definition of an admissible square are easily checked.

6.7.6. LEMMA. Given an admissible square of depth n

$$(S) \qquad \begin{array}{ccc} \underline{\underline{B}} & \overset{q}{\rightsquigarrow} & \underline{\underline{N}} \\ {\scriptstyle f}\downarrow & & \downarrow{\scriptstyle g} \\ \underline{\underline{A}} & \underset{p}{\longrightarrow} & \underline{\underline{M}} \end{array}$$

and $\varepsilon : \underline{\underline{M}} \dashrightarrow \underline{\underline{R}}^*_+$, consider $\underline{\underline{C}}(q)$ and $\underline{\underline{C}}(p)^\varepsilon$ (see 5.3.6). Define $f_S : C(q) \to C(p)$ by

$$f_S([b,t]) = [f(b),t], \quad b \in B, \quad 0 \leq t \leq 1$$

$$f_S([x,0]) = [g(x),0], \quad x \in N .$$

Then f_S is an a.T.m. from $\underline{\underline{C}}(q)$ to $\underline{\underline{C}}(p)^\varepsilon$.

Proof. Note first that (cf. 5.3.6) $\underline{\underline{C}}(q)$ is a w.a.s. and $\underline{\underline{C}}(p)^\varepsilon$ is an a.s. It is obvious that f_S is a submersive weak morphism. Condition (5.5.2) is a consequence of (viii) in 5.3.6 (since f is an a.T.m. and $\underline{\underline{N}}$ is an a.s.). Condition (5.5.3) follows from the construction of $\underline{\underline{C}}(q)$, using the fact that f is an a.T.m. Finally (5.5.4) is a consequence of (11) in 6.7.1 for strata $Z < Z'$ of $\underline{\underline{C}}(q)$ with $Z \subset i_q(N)$ and $Z' \not\subset i_q(N)$, of 5.9.6, 5.3.6 (ii) and of the fact that f is an a.T.m. for strata $Z < Z'$ of $\underline{\underline{C}}(q)$ with $Z \subset i_q(N)$, and of 5.9.4 and 5.3.6(v) for strata $Z < Z'$ of $\underline{\underline{C}}(q)$ with $Z' \subset i_q(N)$. Q.E.D.

6.7.6. Let S be an admissible square of depth n. Besides f_S we can also consider the a.T.m.'s $f_{S^+} : \underline{\underline{C}}(q^+) \vdash\!\!\rightarrow \underline{\underline{C}}(p)^\varepsilon$ and $f_{S_e} : \underline{\underline{C}}(q_e) \vdash\!\!\rightarrow \underline{\underline{C}}(p)^\varepsilon$ defined in the same way as f. Let $q^0 = q|B^0 : B^0 \to N^0$. Then $C(q^0)$ is a face of both $\underline{\underline{C}}(q^+)$ and $\underline{\underline{C}}(q_e)$ and we can consider $\underline{\underline{C}}(q^+) \cup_{C(q^0)} \underline{\underline{C}}(q_e)$.

In view of 5.9.8, we can also consider the a.T.m.

$\overline{f} = f_{S^+} \cup_{C(q^0)} f_{S_e} : \underline{\underline{C}}(q^+) \cup_{C(q^0)} \underline{\underline{C}}(q_e) \vdash\!\!\rightarrow \underline{\underline{C}}(p)^\varepsilon$. Let also

$G : \underline{\underline{C}}(q^+) \cup_{C(q^0)} \underline{\underline{C}}(q_e) \dashrightarrow \underline{\underline{C}}(q)$ be the weak isomorphism constructed in 5.3.8 (notice that $q = q^+ \cup_{B^0} q_e$). A direct verification shows that $(G, 1_{C(p)})$ is an isomorphism from $\overline{f}$ to f_S. In particular, the diagram

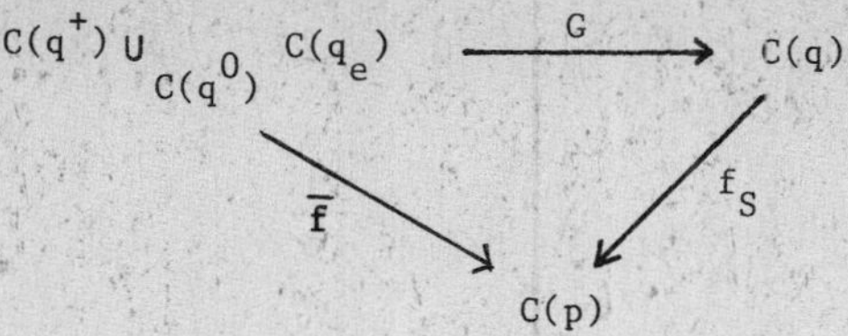

is commutative.

6.7.7. Later on certain constructions will lead us to consider squares

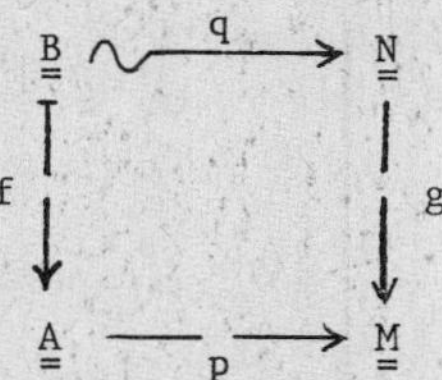

in which N and B, or A and B, or B are the empty sets. Such a square will also be called admissible of depth $n = \text{depth}(\underline{\underline{N}})$ ($\text{depth}(\emptyset) = -1$ by definition). All the constructions performed until now are still valid.

6.8.1. From now on $f : \underline{\underline{B}} \vdash\!\!\longrightarrow \underline{\underline{A}}$ will be an a.T.m., $\underline{\underline{A}}$ and $\underline{\underline{B}}$ being of finite depth and $\text{depth}(\underline{\underline{A}}) > 0$. Let $B_* = f^{-1}(A^*)$ and $f_* : B_* \to A^*$ be the restriction of f. Since B_* is a closed union of strata, $\underline{\underline{B}}|B_*$ exists; denote it $\underline{\underline{B}}_*$. From Remark (2) in 5.5, it follows that f_* is an a.T.m. from $\underline{\underline{B}}_*$ to $\underline{\underline{A}}^*$. By 5.9.4, $\underline{\underline{B}}_*$ is an a.s. Since $\text{depth}(\underline{\underline{B}}_*) \leq \text{depth}(\underline{\underline{B}}) < \infty$, we can consider $(\underline{\underline{B}}_*)^*$, denoted $\underline{\underline{B}}_*^*$. If $B_* \neq \emptyset$, $\text{depth}(\underline{\underline{B}}_*^*) = 0$; if so let $f_*^* : \underline{\underline{B}}_*^* \longrightarrow \underline{\underline{A}}^*$ be the restriction of f and choose $\delta : \underline{\underline{B}}_*^* \longrightarrow \underline{\underline{R}}_+^*$ and $\varepsilon : \underline{\underline{A}}^* \longrightarrow \underline{\underline{R}}_+^*$ such that all the conditions involved hold on $T_Y^{\delta|Y}$ and $T_X^{\varepsilon|X}$, $Y \in \mathcal{B}_*^*$, $X \in \mathcal{A}^*$. By (5.1.9), $T_X^{\varepsilon|X} \cap T_{X'}^{\varepsilon|X'} = \emptyset$ if $X \neq X'$ are strata of $\underline{\underline{A}}^*$ and $T_Y^{\delta|Y} \cap T_{Y'}^{\delta|Y'} = \emptyset$ if $Y \neq Y'$ are strata of $\underline{\underline{B}}_*^*$. Let $A^{\varepsilon)} = \bigcup_{X \in \mathcal{A}^*} T_X^{\varepsilon|X}$ and $B_{\varepsilon)}^{\delta)} = (\bigcup_{Y \in \mathcal{B}_*^*} T_Y^{\delta|Y}) \cap f^{-1}(A^{\varepsilon)})$ and define $\pi^{\varepsilon)} : A^{\varepsilon)} \to A^*$ and $\pi_{\varepsilon)}^{\delta)} : B_{\varepsilon)}^{\delta)} \to B_*^*$ by $\pi^{\varepsilon)}|T_X^{\varepsilon|X} = \pi_X|T_X^{\varepsilon|X}$ and $\pi_{\varepsilon)}^{\delta)}|T_Y^{\delta|Y} \cap f^{-1}(A^{\varepsilon)}) = \pi_Y|T_Y^{\delta|Y} \cap f^{-1}(A^{\varepsilon)})$. It is obvious that $\pi^{\varepsilon)}$ (resp. $\pi_{\varepsilon)}^{\delta)}$) is a submersive weak morphism from $\underline{\underline{A}}^{\varepsilon)} = \underline{\underline{A}}|A^{\varepsilon)}$ (resp.

$\underline{\underline{B}}^{\delta)}_{\varepsilon)} = \underline{\underline{B}}|B^{\delta)}_{\varepsilon)}$ to $\underline{\underline{A}}^*$ (resp. $\underline{\underline{B}}_*^*$). Consider the fibre product $\underline{\underline{B}}_*^* \times_{A^*} \underline{\underline{A}}^{\varepsilon)}$ with respect to f_*^* and $\pi^{\varepsilon)}$ (see 5.9.3). Since $f_*^* \circ \pi^{\delta)}_{\varepsilon)} = \pi^{\varepsilon)} \circ (f|B^{\delta)}_{\varepsilon)})$ we can define a map $g : B^{\delta)}_{\varepsilon)} \to B_*^* \times_{A^*} A^{\varepsilon)}$ by $g(b) = (\pi^{\delta)}_{\varepsilon)}(b), f(b))$. Locally this map is just the map constructed in 5.9.5 and therefore (cf. Remark (1) in 5.5) g is an a.T.m. from $\underline{\underline{B}}^{\delta)}_{\varepsilon)}$ to $\underline{\underline{B}}_*^* \times_{A^*} \underline{\underline{A}}^{\varepsilon)}$.

6.8.2. A <u>decomposition</u> of f is a quintuple $\nabla = \{\Delta, \underline{\underline{B}}_-, \underline{\underline{B}}_+, S, \Psi\}$, where

(i) $\Delta = \{\underline{\underline{A}}^-, \underline{\underline{A}}^+, p, \varepsilon, \Phi\}$ is a decomposition of $\underline{\underline{A}}$;

(ii) $\underline{\underline{B}}_-$ (resp. $\underline{\underline{B}}_+$) is a w.a.s. structure on $B_- = f^{-1}(A^-)$ (resp. $B_+ = f^{-1}(A^+)$), $f_- = f|B_- : \underline{\underline{B}}_- \longrightarrow \underline{\underline{A}}^-$ (resp. $f_+ = f|B_+ : \underline{\underline{B}}_+ \longrightarrow \underline{\underline{A}}^+$) is an a.T.m., $B_0 = B_- \cap B_+ = f^{-1}(A^0)$ is an f_--vertical (resp. f_+-vertical) face of $\underline{\underline{B}}_-$ (resp. $\underline{\underline{B}}_+$), $\underline{\underline{B}}_-|B_0 = \underline{\underline{B}}_+|B_0$, one can construct $\underline{\underline{B}}_- \cup_{B_0} \underline{\underline{B}}_+$ and the inclusions $B_- \subset B$, $B_+ \subset B$, $A_- \subset A$ and $A_+ \subset A$ determine an isomorphism from the a.T.m. $f_- \cup_{B_0} f_+ : \underline{\underline{B}}_- \cup_{B_0} \underline{\underline{B}}_+ \vdash\!\!\longrightarrow \underline{\underline{A}}^- \cup_{A^0} \underline{\underline{A}}^+$ to the a.T.m. $f : \underline{\underline{B}} \vdash\!\!\longrightarrow \underline{\underline{A}}$ (as a consequence $\underline{\underline{B}}|B_0$ exists and $f_0 = f|B_0 : \underline{\underline{B}}_0 = \underline{\underline{B}}|B_0 \longrightarrow \underline{\underline{A}}^0$ is an a.T.m.);

(iii) S is an admissible square of depth $n = \text{depth}(\underline{\underline{B}}_*)$, namely

(S)

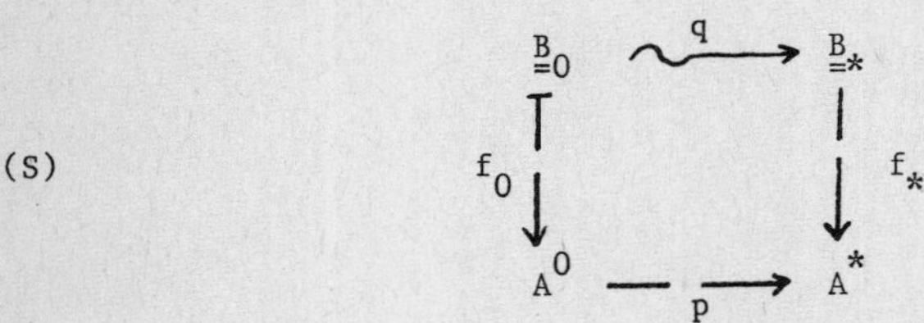

(iv) (Ψ,Φ) is an isomorphism from the a.T.m. $f_S : \underline{\underline{C}}(q) \vdash\!\!\longrightarrow \underline{\underline{C}}(p)^{\varepsilon}$ to the a.T.m. $f_- : \underline{\underline{B}} \vdash\!\!\longrightarrow \underline{\underline{A}}^-$ such that $\Psi([b,0]) = b$ if $b \in B_*$ and $\Psi([b,1]) = b$ if $b \in B_0$.

<u>Remarks</u>. (1) If $\text{depth}(\underline{\underline{A}}) = 0$, giving a decomposition $\nabla = \{\Delta, \underline{\underline{B}}_-, \underline{\underline{B}}_+, S, \Psi\}$ of f is equivalent to giving the regular and f-compatible total decomposition $\mathcal{D}_S$ of $\underline{\underline{B}} = \underline{\underline{B}}_*$. If $B_* = \emptyset$ and $\nabla = \{\Delta, \underline{\underline{B}}_-, \underline{\underline{B}}_+, S, \Psi\}$ is a decomposition of f, then $B_- = \emptyset = B_0$, $\underline{\underline{B}}_+ = \underline{\underline{B}}$ and ∇ is completely determined by Δ.

(2) Let ∇ be a decomposition of f. Since depth($\underline{\underline{A}}_+$) = depth($\underline{\underline{A}}$) - 1, $f_+: \underline{\underline{B}}_+ \vdash\!\rightarrow \underline{\underline{A}}^+$ is "simpler" than f. On the other hand, since f_S is completely determined by S and since depth($\underline{\underline{A}}^0$) = depth($\underline{\underline{A}}$) - 1 and depth($\underline{\underline{A}}^*$) = 0, f_S is "simpler" than f. If depth($\underline{\underline{B}}_*$) $\geq$ 1 notice that f_S can be decomposed into f_{S^+} and f_{S_e} (see 6.7.6) which in turn are "simpler" than f_S.

6.8.3. Let $f : \underline{\underline{B}} \vdash\!\rightarrow \underline{\underline{A}}$ and $\tilde{f} : \underline{\underline{\tilde{B}}} \vdash\!\rightarrow \underline{\underline{\tilde{A}}}$ be a.T.m. 's, $\nabla = \{\Delta, \underline{\underline{B}}_-, \underline{\underline{B}}_+, S, \Psi\}$ and $\tilde{\nabla} = \{\tilde{\Delta}, \underline{\underline{\tilde{B}}}_-, \underline{\underline{\tilde{B}}}_+, \tilde{S}, \tilde{\Psi}\}$ be decompositions of f and $\tilde{f}$ respectively and (G,F) be an isomorphism from f to $\tilde{f}$. (G,F) is called <u>compatible with</u> ∇ <u>and</u> $\tilde{\nabla}$ if

(i) F is compatible with Δ and $\tilde{\Delta}$ (see 6.1.2);

(ii) if $G_- : B_- \rightarrow \tilde{B}_-$ denotes the restriction of G, then (G_-, F^-) is an isomorphism from f_- to $\tilde{f}_-$;

(iii) if $G_+ : B_+ \rightarrow \tilde{B}_+$ denotes the restriction of G, then (G_+, F^+) is an isomorphism from f_+ to $\tilde{f}_+$;

(iv) let $G_0 : B_0 \rightarrow \tilde{B}_0$, $G_* : B_* \rightarrow \tilde{B}_*$, $F^0 : A^0 \rightarrow \tilde{A}^0$ and $F^* : A^* \rightarrow \tilde{A}^*$ be the restrictions of G and F respectively; then (G_0, G_*, F^0, F^*) is an admissible isomorphism from S to $\tilde{S}$;

(v) $G(\Psi([b,t])) = \tilde{\Psi}([G(b),t])$, $b \in B_0$, $0 \leq t \leq 1$.

Two decompositions ∇ and $\tilde{\nabla}$ of f are called <u>equal</u>, denoted $\nabla = \tilde{\nabla}$, if $(1_B, 1_A)$ is compatible with ∇ and $\tilde{\nabla}$.

Let again $f : \underline{\underline{B}} \vdash\!\rightarrow \underline{\underline{A}}$ and $\tilde{f} : \underline{\underline{B}} \vdash\!\rightarrow \underline{\underline{\tilde{A}}}$ be a.T.m. 's, ∇ be a decomposition of f and (G,F) be an isomorphism from f to $\tilde{f}$. It is easy to see that there exists a unique decomposition of $\tilde{f}$, denoted $G_*\nabla$, such that (G,F) is compatible with ∇ and $G_*\nabla$.

6.8.4. Let $\nabla = \{\nabla, \underline{\underline{B}}_-, \underline{\underline{B}}_+, S, \Psi\}$ be a decomposition of f. Let also $B_i \neq \emptyset$ be a face of $\underline{\underline{B}}$ and $f_i : \underline{\underline{B}}_i \vdash\!\rightarrow \underline{\underline{A}}_i$ be the restriction of f (recall that $\underline{\underline{A}}_i = \underline{\underline{A}}$ if B_i is f-horizontal). Suppose that

$\mathrm{depth}(\underline{\underline{A}}_i) = \mathrm{depth}(\underline{\underline{A}})$. We shall define a decomposition $\nabla|B_i$ of f_i as follows. Notice first that we can consider $\Delta|A_i$. Next note that $(B_i)_* = f_i^{-1}(A_i \cap A^*) = B_i \cap B_*$ is a face of $\underline{\underline{B}}_*$, $(B_i)_- = B_- \cap B_i$ is a face of $\underline{\underline{B}}_-$, $(B_i)_\times = B_\times \cap B_i$ is a face of $\underline{\underline{B}}_\times$, $(B_i)_+ = B_+ \cap B_i$ is a face of $\underline{\underline{B}}_+$ and $(B_i)_0 = (B_i)_+ \cap (B_i)_- = B_0 \cap B_i$ is a face of $\underline{\underline{B}}_0$. If $q : \underline{\underline{B}}_0 \rightsquigarrow \underline{\underline{B}}_*$ is the quasimorphism of the square S, then $q^{-1}((B_i)_*) = (B_i)_0$ and thus $S|(B_i)_0$ is the square

$$\begin{array}{ccc} \underline{\underline{B}}_0|(B_i)_0 & \overset{q_i}{\rightsquigarrow} & B_*|(B_i)_* \\ {\scriptstyle f_{0,i}}\downarrow & & \downarrow{\scriptstyle f_{*,i}} \\ \underline{\underline{A}}^0|A^0 \cap A_i & \underset{p_i}{\longrightarrow} & \underline{\underline{A}}^*|A^* \cap A_i \end{array}$$

the mappings being restrictions of the corresponding mappings in S. It is obvious that $\Psi_i = \Psi|C(q_i) : \underline{\underline{C}}(q_i) \longrightarrow \underline{\underline{B}}_-|(B_i)_-$ is a weak isomorphism. Finally, a straightforward verification shows that $\nabla|B_i = \{\Delta|A_i, \underline{\underline{B}}_-|(B_i)_-, \underline{\underline{B}}_+|(B_i)_+, S|(B_i)_0, \Psi_i\}$ is a decomposition of the a.T.m. $f_i : \underline{\underline{B}}_i \longmapsto \underline{\underline{A}}_i$.

6.8.5. To a decomposition $\nabla = \{\Delta, \underline{\underline{B}}_-, \underline{\underline{B}}_+, S, \Psi\}$ of f one can associate two weakly controlled vector fields ξ_∇ and ζ_∇ which will play an important role in what follows. Before defining them, let us fix (and recall) some more notation.

S being the square

$$\begin{array}{ccc} \underline{\underline{B}}_0 & \overset{q}{\rightsquigarrow} & \underline{\underline{B}}_* \\ {\scriptstyle f_0}\downarrow & & \downarrow{\scriptstyle f_*} \\ \underline{\underline{A}}^0 & \underset{p}{\longrightarrow} & \underline{\underline{A}}^* \end{array}$$

let $\Delta_S = \{\underline{\underline{B}}_*, \underline{\underline{B}}_*^+, p_*, \delta, \Phi_*\}$ be the associated decomposition of $\underline{\underline{B}}_*$, let

$$(S^+)\qquad \begin{array}{ccc} \underline{\underline{B}}_0^+ & \overset{q}{\rightsquigarrow} & \underline{\underline{B}}_*^+ \\ {\scriptstyle f_0^+}\downarrow & & \downarrow{\scriptstyle f_*^+} \\ \underline{\underline{A}}^0 & \xrightarrow[p]{} & A^* \end{array}$$

be the admissible square of depth $n-1$ associated to S $(n = \operatorname{depth}(\underline{\underline{B}}_*))$ and let

$$(S^-)\qquad \begin{array}{ccc} \underline{\underline{B}}_0^- & \xrightarrow{q^-} & \underline{\underline{B}}_*^* \\ {\scriptstyle f_0^*}\downarrow & & \downarrow{\scriptstyle f_*^*} \\ \underline{\underline{A}}^0 & \xrightarrow[p]{} & \underline{\underline{A}}^* \end{array}$$

be the basic square associated to S. Let also $\underline{\underline{B}}_C^\times$, $B_0^h = B_0^- \cap B_0^\times$, $\theta_S : \underline{\underline{B}}_0^h \times [0,1] \dashrightarrow \underline{\underline{B}}_0^\times$, $B_0^0 = B_0^\times \cap B_0^+$ and $q^0 : B_0^0 \to B_*^0 = B_*^+ \cap B_*^-$ be the other data associated to S.

Let next $F_{B_0}^+ : U_{B_0}^+ \to B_0 \times R_+$, $p_{B_0}^+ : U_{B_0}^+ \to B_0$, $r_{B_0}^+ : U_{B_0}^+ \to R_+$ and $\eta_{B_0}^+$ be the data associated to the face B_0 of $\underline{\underline{B}}_+$ and $F_{A^0}^+ : U_{A^0}^+ \to A^0 \times R_+$, $p^+ : U_{A^0}^+ \to A^0$, $r_{A^0}^+ : U_{A^0}^+ \to R_+$ and $\eta_{A^0}^+$ be the data associated to the face A^0 of $\underline{\underline{A}}^+$. We can (and shall) assume that

$$U_{B_0}^+ \subset f^{-1}(U_{A^0}^+) ,$$

$$df \cdot \eta_{B_0}^+ = \eta_{A^0}^+$$

and the diagram

$$\begin{array}{ccc} U_{B_0}^+ & \xrightarrow{F_{B_0}^+} & B_0 \times R_+ \\ {\scriptstyle f|U_{B_0}^+}\downarrow & & \downarrow{\scriptstyle f_0 \times 1_{R_+}} \\ U_{A_0}^+ & \xrightarrow[F_{A^0}^+]{} & A^0 \times R_+ \end{array}$$

is commutative. If in addition f is proper, we can (and shall) assume that

$$U^+_{B_0} = f^{-1}(U^+_{A_0}).$$

Set $U_\nabla = B_- \cup U^+_{B_0}$ and define $\varphi_\nabla : U_\nabla \to R_+$, $\pi_\nabla : U_\nabla \to B_*$ and $\xi_\nabla \in \mathfrak{X}^f_{\underline{\underline{B}}}(U_\nabla \setminus B_*)$ by

$$\varphi_\nabla(\Psi([b,t])) = t, \quad [b,t] \in C(q),$$

$$\varphi_\nabla(b) = 1 + r^+_{B_0}(b), \quad b \in U^+_{B_0},$$

$$\pi_\nabla(\Psi([b,t])) = \pi_q([b,t]), \quad [b,t] \in C(q),$$

$$\pi_\nabla(b) = q(p^+_{B_0}(b)), \quad b \in U^+_{B_0},$$

$$\xi_\nabla | B_- \setminus B_* = d(\Psi \circ \varphi_q) \cdot (0 \times d/dt),$$

$$\xi_\nabla | U^+_{B_0} = \eta^+_{B_0}$$

($\pi_q : C(q) \to B_*$ and $\varphi_q : B_0 \times [0,1] \to C(q)$ are defined in 5.3.6). It is clear that

$$df \cdot \xi_\nabla = \xi_\Delta \; ; \quad \varphi_\nabla = \varphi_\Delta \circ (f | U_\nabla);$$

$$\{(b,t) \in (U_\nabla \setminus B_*) \times R; -\varphi_\nabla(b) < t \leq 0\} = \{(b,t) \in D_{\xi_\nabla} ; \; t \leq 0\}$$

$$\Psi([b,t]) = \lambda_{\xi_\nabla}(b,t-1), \quad b \in B_0, \quad 0 < t \leq 1 \; ;$$

$$\lim_{t \nearrow 1} \lambda_{\xi_\nabla}(b,-t) = q(b), \quad b \in B_0 \; ;$$

$$F^+_{B_0}(b) = (\lambda_{\xi_\nabla}(b,1-\varphi_\nabla(b)), \; \varphi_\nabla(b)-1), \quad b \in U^+_{B_0} \; ;$$

$$\pi_\nabla(\lambda_{\xi_\nabla}(b,t)) = \pi_\nabla(b), \quad (b,t) \in D_{\xi_\nabla} \; ;$$

$$\varphi_\nabla(\lambda_{\xi_\nabla}(b,t)) = \varphi_\nabla(b) + t, \quad (b,t) \in D_{\xi_\nabla} .$$

In a certain sense ξ_∇ determines the "horizontal" structure of $\underline{\underline{B}}$ near B_*. However, in general, this is not enough to determine the structure of $\underline{\underline{B}}$ near B_* : $\underline{\underline{B}}$ has also a certain "vertical" structure. Part of this structure will be determined by the vector field ζ_∇ which we proceed now to construct.

Let $F^-_{B_0^h} : U^-_{B_0^h} \to B_0^h \times R_+$ and $F^+_{B_0^0} : U^+_{B_0^0} \to B_0^0 \times R_+$ be the collars of B_0^h in B_0^- and B_0^0 in B_0^+ respectively. Let $\eta^-_{B_0^h}$ and $\eta^+_{B_0^0}$ be the associated vector fields. Set

$$W_0 = U^-_{B_0^h} \cup B_0^\times \cup U^+_{B_0^0} \subset B_0$$

and define $\zeta_0 \in \mathcal{X}^{f_0}_{\underline{\underline{B}}_0}(W_0)$ by

$$\zeta_0 | U^-_{B_0^h} = -\eta^-_{B_0^h}$$

$$\zeta_0 | B_0^\times = d\theta_S \cdot (0 \times d/dt) \ ,$$

$$\zeta_0 | U^+_{B_0^0} = \eta^+_{B_0^0} \ .$$

Next let $W_+ = \{b \in U^+_{B_0} \ ; \ p^+_{B_0}(b) \in W_0\}$ and let $\zeta_+ \in \mathcal{X}^{f_+}_{\underline{\underline{B}}_+}(W_+)$ be determined by

$$dp^+_{B_0} \cdot \zeta_+ = \zeta_0$$

and

$$dr^+_{B_0} \cdot \zeta_+ = 0 \ .$$

Since q is proper, we may assume that

$$q^{-1}(U^+_{B_*^0}) = U^+_{B_0^0}$$

and the diagram

$$\begin{array}{ccc} U^{+}_{B^0_0} & \xrightarrow{q|U^{+}_{B^0_0}} & U^{+}_{B^0_*} \\ F^{+}_{B^0_0}\downarrow & & \downarrow F^{+}_{B^0_*} \\ B^0_0 \times R_+ & \xrightarrow[q^0 \times 1_{R_+}]{} & B^0_* \times R_+ \end{array}$$

is commutative (the right vertical map is the collar of B^0_* in B^+_*). Set

$$W_- = (U_{\Delta_S} \setminus B^*_*) \cup \Psi(\varphi_q(W_0 \times (0,1]))$$

and

$$W_\nabla = W_- \cup W_+ .$$

It is obvious that W_∇ is open B. Define $\zeta_\nabla \in X^f_{\underline{\underline{B}}}(W_\nabla)$ by

$$\zeta_\nabla | U_{\Delta_S} \setminus B^*_* = \xi_{\Delta_S} ,$$

$$\zeta_\nabla | \Psi(\varphi_q(W_0 \times (0,1])) = d(\Psi \circ \varphi_q) \cdot (\zeta_0 \times 0),$$

$$\zeta_\nabla | W_+ = \zeta_+ .$$

From the definitions it follows immediately that

$$df \cdot \zeta_\nabla = 0$$

and

$$[\zeta_\nabla, \xi_\nabla] = 0 .$$

We shall also define maps $\psi_\nabla : W_\nabla \to R$ and $\hat{\pi}_\nabla : W_\nabla \to B^*_*$ by

$\psi_\nabla(b) = -r^-_{B^h_0}(b)$, $\hat{\pi}_\nabla(b) = q(b)$ if $b \in U^-_{B^h_0}$;

$\psi_\nabla(b) = t$, $\hat{\pi}_\nabla(b) = q(b_0)$ if $b = \theta_S(b_0,t) \in B^\times_0$, $b_0 \in B^h_0$, $0 \le t \le 1$;

$\psi_\nabla(b) = 1 + r^+_{B^0_0}(b)$, $\hat{\pi}_\nabla(b) = \pi_{\Delta_S}(q(b))$, if $b \in U^+_{B^0_0}$;

$\psi_\nabla(b) = \psi_\nabla(p^+_{B_0}(b))$, $\hat{\pi}_\nabla(b) = \hat{\pi}_\nabla(p^+_{B_0}(b))$ if $b \in W_+$ (the right sides are already defined;

$\psi_\nabla(b) = \psi_\nabla(b_0)$, $\hat{\pi}_\nabla(b) = \hat{\pi}_\nabla(b_0)$ if $b = \Psi([b_0,t]) \in W_-$, $b_0 \in W_0$, $0 < t \leq 1$ (the right sides are already defined);

$\psi_\nabla(b) = \varphi_{\Delta_S}(b)$, $\hat{\pi}_\nabla(b) = \pi_{\Delta_S}(b)$ if $b \in U_{\Delta_S} \setminus B^*_*$.

Notice that ψ_∇ is controlled, $\hat{\pi}_\nabla$ is a submersive weak morphism from $\underline{\underline{B}}|W_\nabla$ to $\underline{\underline{B}}^*_*$ and

$d\psi_\nabla \cdot \zeta_\nabla = d/dt$, $d\psi_\nabla \cdot \xi_\nabla = 0$,

$d\hat{\pi}_\nabla \cdot \zeta_\nabla = 0$, $d\hat{\pi}_\nabla \cdot \xi_\nabla = 0$,

$\hat{\pi}_\nabla = \pi_{\Delta_S} \circ (\pi_\nabla | W_\nabla)$.

For later use set

$$B^+_- = \Psi(C(q^+)),\quad B^\times_- = \Psi(C(q^\times)),\quad B^-_- = \Psi(C(q^-)).$$

6.8.6. Let $\nabla = \{\Delta, \underline{\underline{B}}_-, \underline{\underline{B}}_+, S, \Psi\}$ and $\tilde{\nabla} = \{\tilde{\Delta}, \tilde{\underline{\underline{B}}}_-, \tilde{\underline{\underline{B}}}_+, \tilde{S}, \tilde{\Psi}\}$ be decompositions of f. It is obvious that if $\nabla = \tilde{\nabla}$ then $B_- = \tilde{B}_-$, $B^-_- = \tilde{B}^-_-$, $U_\nabla = U_{\tilde{\nabla}}$ near B_-, $\xi_\nabla = \xi_{\tilde{\nabla}}$ near B_-, $W_\nabla = W_{\tilde{\nabla}}$ near B^-_- and $\zeta_\nabla = \zeta_{\tilde{\nabla}}$ near B^-_-.

The converse is not true because the above equalities do not imply $S \equiv \tilde{S}$ (however, they imply that $\Delta = \tilde{\Delta}$, $\underline{\underline{B}}_- = \tilde{\underline{\underline{B}}}_-$, $\underline{\underline{B}}_+ = \tilde{\underline{\underline{B}}}_+$ and $\Psi = \tilde{\Psi}$; thus, if in addition $S \equiv \tilde{S}$, then $\nabla = \tilde{\nabla}$).

6.8.7. Let $\Delta = \{\underline{\underline{A}}^-, \underline{\underline{A}}^+, p, \varepsilon, \Phi\}$ be a decomposition of $\underline{\underline{A}}$ and $\nabla = \{\Delta, \underline{\underline{B}}_-, \underline{\underline{B}}_+, S, \Psi\}$ be a decomposition of f. Let U be an open subset of A containing A^- and let V be an open subset of B containing B_- and contained in $f^{-1}(U)$. Set $f_{V,U} = f|V : \underline{\underline{B}}|V \longmapsto \underline{\underline{A}}|U$. Then $\nabla_{V,U} = \{\Delta|U, \underline{\underline{B}}_-, \underline{\underline{B}}_+|B_+ \cap V, S, \Psi\}$ is a decomposition of $f_{V,U}$.

6.8.8. Given a decomposition $\nabla = \{\Delta, \underline{\underline{B}}_-, \underline{\underline{B}}_+, S, \Psi\}$ of f, set

$B^+_- = \Psi(C(q^+))$, $B^+_+ = (p^+_{B_0})^{-1}(B^+_0)$, $B^+ = B^+_+ \cup B^+_-$, $f^+ = f|B^+ : B^+ \to U_\Delta$ and $\Psi^+ = \Psi|C(q^+) : C(q^+) \to B^+_-$. One can endow B^+_-, B^+_+ and B^+ with canonical w.a.s. structures $\underline{\underline{B}}^+_-$, $\underline{\underline{B}}^+_+$ and $\underline{\underline{B}}^+$ respectively such that f^+ is an a.T.m. from $\underline{\underline{B}}^+$ to $\underline{\underline{A}}|U_\Delta$ and $\nabla^+ = \{\Delta|U_\Delta, \underline{\underline{B}}^+_-, \underline{\underline{B}}^+_+, S^+, \Psi^+\}$ is a decomposition of f^+.

Consider also $B^0_- = \Psi(C(q^0))$, $B^0_+ = (p^+_{B_0})^{-1}(B^0_0)$, $B^0 = B^0_+ \cup B^0_-$, $f^0 = f|B^0 : B^0 \to U_\Delta$ and $\Psi^0 = \Psi|C(q^0) : C(q^0) \to B^0_-$ (recall that $q^0 = q|B^0_0 : B^0_0 \to B^0_*$). Notice that B^0 is a face of $\underline{\underline{B}}^+$ and $B^0_+ = B^+_+ \cap B^0$, $B^0_- = B^+_- \cap B^0$ and $\Psi^0 = \Psi^+|C(q^0)$. We can therefore consider the decomposition $\nabla^+|B^0$ of f^0. We shall denote it ∇^0 ; it is obvious that ∇^0 is of the form $\nabla^0 = \{\Delta|U_\Delta, \underline{\underline{B}}^0_-, \underline{\underline{B}}^0_+, S^0, \Psi^0\}$, where $S^0 = S^+|B^0_0$.

Let now $\tilde{A} = (\underline{\underline{A}}|U_\Delta) \times_{A^*} \underline{\underline{B}}^*_*$, the fibre product being taken with respect to π_Δ and f^*_* , and let $\tilde{f}^0 : B^0 \to \tilde{A}$ be given by $\tilde{f}^0(b) = (f^0(b), \hat{\pi}_\nabla(b))$. A direct verification shows that $\tilde{f}^0$ is an a.T.m. from $\underline{\underline{B}}^0$ to $\underline{\underline{\tilde{A}}}$. The decomposition Δ of $\underline{\underline{A}}$ determines a decomposition $\tilde{\Delta} = \{\underline{\underline{\tilde{A}}}^-, \underline{\underline{\tilde{A}}}^+, \tilde{p}, \tilde{\varepsilon}, \tilde{\Phi}\}$ of $\underline{\underline{\tilde{A}}}$ as in 6.1.10. In view of (12) in 6.7.1 the square

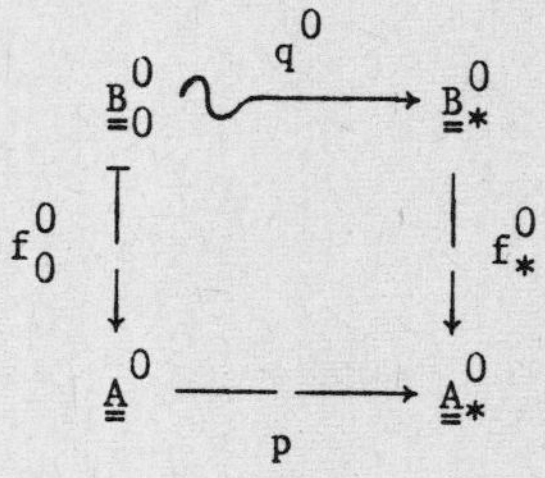

$$\begin{array}{ccc} \underline{\underline{B}}^0_0 & \overset{q^0}{\rightsquigarrow} & \underline{\underline{B}}^0_* \\ {\scriptstyle f^0_0}\downarrow & & \downarrow{\scriptstyle f^0_*} \\ \underline{\underline{A}}^0 & \underset{p}{\longrightarrow} & \underline{\underline{A}}^0_* \end{array}$$

determines an admissible square

$$(\tilde{S}^0) \qquad \begin{array}{ccc} \underline{\underline{B}}^0_0 & \overset{q^0}{\rightsquigarrow} & \underline{\underline{B}}^0_* \\ {\scriptstyle \tilde{f}|B^0_0}\downarrow & & \searrow{\scriptstyle p_*} \\ \underline{\underline{\tilde{A}}}^0 & \underset{\tilde{p}}{\longrightarrow} & \underline{\underline{\tilde{A}}}^* = \underline{\underline{B}}^*_* \end{array}$$

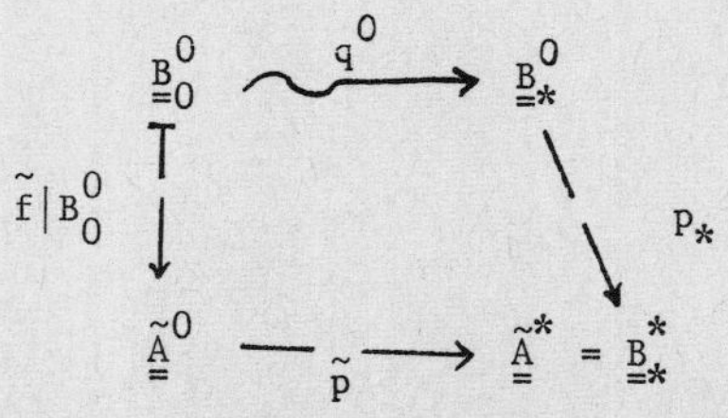

where the identification of $\underline{\underline{\tilde{A}}}^* = \underline{\underline{A}}^* \times_{A^*} \underline{\underline{B}}^*_*$ with $\underline{\underline{B}}^*_*$ is given by

$(a,b) \mapsto b$ and p_* is the weak morphism involved in the decomposition Δ_S of $\underline{\underline{B}}_*$ (in order to apply (12) of 6.7.1 we have to use (7) of the same subsection). Now it is easy to check that $\tilde{\nabla}^0 = \{\tilde{\Delta}, \underline{\underline{B}}_-^0, \underline{\underline{B}}_+^0, \tilde{S}^0, \Psi^0\}$ is a decomposition of $\tilde{f}^0$.

6.8.9. Assume now that $f : \underline{\underline{B}} \mapsto\!\!\to \underline{\underline{A}}$ is proper. Let $\nabla = \{\Delta, \underline{\underline{B}}_-, \underline{\underline{B}}_+, S, \Psi\}$ be a decomposition of f, let $\Delta = \{\underline{\underline{A}}^-, \underline{\underline{A}}^+, p, \varepsilon, \Phi\}$ and let $\mu : \underline{\underline{A}}^* \dashrightarrow \underline{\underline{R}}|(0,1)$. Since f is proper, we may assume that $U_\nabla = f^{-1}(U_\Delta)$ and $df \cdot \xi_\nabla = \xi_\Delta$. Consider the decomposition $\Delta_\mu = \{\tilde{\underline{\underline{A}}}_-, \tilde{\underline{\underline{A}}}_+, \tilde{p}, \tilde{\varepsilon}, \tilde{\Phi}\}$ of $\underline{\underline{A}}$ (see 6.1.7). We shall define a decomposition $\nabla_\mu = \{\Delta_\mu, \tilde{\underline{\underline{B}}}_-, \tilde{\underline{\underline{B}}}_+, \tilde{S}, \tilde{\Psi}\}$ of f as follows. By definition $\tilde{B}_- = f^{-1}(\tilde{A}^-)$ and $\tilde{B}_+ = f^{-1}(\tilde{A}^+)$; thus $\tilde{B}_0 = f^{-1}(\tilde{A}^0)$. Set $\xi = (\mu \circ \pi_\Delta \circ f)\xi_\nabla \in \mathcal{X}^f_{\underline{\underline{B}}}(U_\nabla\ B_*)$; then $df \cdot \xi = (\mu \circ \pi_\Delta)\xi_\Delta = \xi_{\Delta_\mu}$. We can endow $\tilde{B}_-$ (resp. $\tilde{B}_+$) with a w.a.s. structure $\tilde{\underline{\underline{B}}}_-$ (resp. $\tilde{\underline{\underline{B}}}_+$) which verifies 6.8.2(ii) and such that the vector field associated to its face $\tilde{B}_0$ is a restriction of ξ. Next define $G : B_0 \to \tilde{B}_0$ and $F: A^0 \to \tilde{A}^0$ by setting

$$G(b) = \lambda_{\xi_\nabla}(b, \mu(\pi_\Delta(f(b))) - 1),$$

$$F(a) = \lambda_{\xi_\Delta}(a, \mu(\pi_\Delta(a)) - 1)$$

and notice that in the commutative diagram

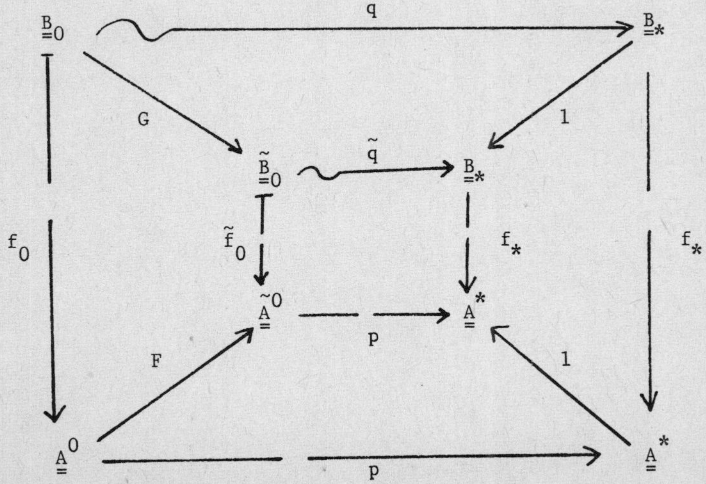

(G,F) is an isomorphism from the a.T.m. f_0 to the a.T.m. $\tilde{f}_0$. By 6.7.2, we can endow $\tilde{S}$, the interior square, with a structure of an admissible square. Finally $\tilde{\Psi}$ will be determined by the condition $\xi_{\nabla_\mu} = \xi$.

It is useful to observe that we can take $U_{\nabla_\mu} = U_\nabla$, $W_{\nabla_\mu} = W_\nabla$ and then $\xi_{\nabla_\mu} = (\mu \circ \pi_\Delta \circ f)\xi_\nabla$, $\pi_{\nabla_\mu} = \pi_\nabla$, $\varphi_{\nabla_\mu} = (1/\mu \circ \pi_\Delta \circ f)\varphi_\nabla$, $\zeta_{\nabla_\mu} = \zeta_\nabla$, $\psi_{\nabla_\mu} = \psi_\nabla$.

6.8.10. Assume again that $f : \underline{\underline{B}} \longmapsto \underline{\underline{A}}$ is proper, let Δ be a decomposition of $\underline{\underline{A}}$, let $\mu : \underline{\underline{A}}^* \longrightarrow \underline{\underline{R}}|(0,1)$ and let $\tilde{\nabla} = \{\Delta_\mu, \tilde{\underline{\underline{B}}}_-, \tilde{\underline{\underline{B}}}_+, \tilde{S}, \Psi\}$ be a decomposition of f. Later on, we shall be interested in finding a decomposition $\nabla = \{\Delta, \underline{\underline{B}}_-, \underline{\underline{B}}_+, S, \Psi\}$ of f such that $\nabla_\mu = \tilde{\nabla}$. This can be done exactly as above (see 6.8.9) if we can take $U_{\tilde{\nabla}} = f^{-1}(U_{\Delta_\mu})$ and $U_{\Delta_\mu} = U_\Delta$ (in contrast with 6.8.9, we must now "enlarge" $U_{\tilde{\nabla}}$ to equal $f^{-1}(U_\Delta)$). Clearly, we can take $U_{\Delta_\mu} = U_\Delta$. Then using Proposition 5.6 and a partition of unity argument, we can construct a vector field $\xi \in X^f_{\underline{\underline{B}}}(f^{-1}(U_\Delta)\setminus B_*)$ such that $\xi = \xi_{\nabla_\mu}$ near $\tilde{B}_-$ and $df \cdot \xi = \xi_{\Delta_\mu}$. Now we can take the collar of $\tilde{B}_0$ in $\tilde{\underline{\underline{B}}}_+$ to be defined on $U^+_{\tilde{B}_0} = f^{-1}(U_\Delta \setminus \tilde{A}^-) \cup \tilde{B}_0$ and determined by the condition: "its associated vector field is $\xi|U^+_{\tilde{B}_0}$". From the construction of U_{∇_μ} this means that $U_{\nabla_\mu} = f^{-1}(U_\Delta)$.

As a matter of fact, we shall need a slightly more general version of this construction. Namely assume that in addition to our previous hypotheses there is given $I \subset I_{\underline{\underline{B}}}$ and for any $i \in I$ a decomposition $\nabla_i = \{\Delta|A_i, \ldots\}$ of $f_i : \underline{\underline{B}}_i \longmapsto \underline{\underline{A}}_i$ such that $(\nabla_i)_{\mu_i} = \tilde{\nabla}|B_i$ (here $\mu_i = \mu|A_i^*$; notice that, since $\Delta|A_i$ exists, $A_i^* = A_i \cap A^*$). Then we can take ∇ with the additional property that $\nabla|B_i = \nabla_i$, $i \in I$.

6.8.11. Consider the a.T.m. $f : \underline{\underline{B}} \longmapsto \underline{\underline{A}}$ and assume that there exist closed subsets ${}^1B$ and ${}^2B$ of B, endowed with w.a.s. structures ${}^1\underline{\underline{B}}$ and ${}^2\underline{\underline{B}}$ respectively, such that

(1) ${}^{i}f = f|{}^{i}B : {}^{i}\underline{\underline{B}} \longmapsto \underline{\underline{A}}$ is an a.T.m., $i = 1,2$;

(2) ${}_{0}B = {}^{1}B \cap {}^{2}B$ is an ${}^{i}f$-horizontal face of ${}^{i}\underline{\underline{B}}$ and ${}^{1}\underline{\underline{B}}|_{0}B = {}^{2}\underline{\underline{B}}|_{0}B$;

(3) one can construct ${}^{1}\underline{\underline{B}} \cup_{{}_0B} {}^{2}\underline{\underline{B}}$ and the inclusions ${}^{1}B \subset B$ and ${}^{2}B \subset B$ determine an isomorphism from ${}^{1}f \cup_{{}_0B} {}^{2}f : {}^{1}\underline{\underline{B}} \cup_{{}_0B} {}^{2}\underline{\underline{B}} \longmapsto \underline{\underline{A}}$ to $f : \underline{\underline{B}} \longmapsto \underline{\underline{A}}$ over 1_A .

Let also $\nabla_i = \{\Delta, {}^{i}\underline{\underline{B}}_-, {}^{i}\underline{\underline{B}}_+, S_i, \Psi_i\}$ be a decomposition of ${}^{i}f$, $i = 1, 2$, such that $\nabla_1|_{0}B = \nabla_2|_{0}B$. We shall construct a decomposition $\nabla_1 \cup_{{}_0B} \nabla_2 = \{\Delta, \underline{\underline{B}}_-, \underline{\underline{B}}_+, S, \Psi\}$ of f as follows.

First notice that $B_* = {}^{1}B_* \cup {}^{2}B_*$, ${}_{0}B_* = {}_{0}B \cap B_* = {}^{1}B_* \cap {}^{2}B_*$ is a face of both ${}^{1}\underline{\underline{B}}_*$ and ${}^{2}\underline{\underline{B}}_*$ and the inclusions ${}^{1}B_* \subset B_*$ and ${}^{2}B_* \subset B_*$ determine an isomorphism from ${}^{1}f_* \cup_{{}_0B_*} {}^{2}f_* : {}^{1}B_* \cup_{{}_0B_*} {}^{2}B_* \longmapsto \underline{\underline{A}}^*$ to $f_* : \underline{\underline{B}}_* \longmapsto \underline{\underline{A}}^*$ over 1_{A^*} .

Next notice that ${}_{0}B_- = {}^{1}B_- \cap {}^{2}B_-$ is a face of ${}^{i}\underline{\underline{B}}_-$ $(i = 1,2)$, ${}^{1}\underline{\underline{B}}_-|_{0}B_- = {}^{2}\underline{\underline{B}}_-|_{0}B_-$ and we can construct $\underline{\underline{B}}_- = {}^{1}\underline{\underline{B}}_- \cup_{{}_0B_-} {}^{2}\underline{\underline{B}}_-$. By a similar argument, we can construct $\underline{\underline{B}}_+ = {}^{1}\underline{\underline{B}}_+ \cup_{{}_0B_+} {}^{2}\underline{\underline{B}}_+$. Let $B_0 = B_+ \cap B_- = f^{-1}(A^0)$; it is a face of both $\underline{\underline{B}}_-$ and $\underline{\underline{B}}_+$ and $\underline{\underline{B}}_-|B_0 = \underline{\underline{B}}_+|B_0$. Denote this w.a.s. structure on B_0 by $\underline{\underline{B}}_0$. Let $f_0 : \underline{\underline{B}}_0 \longmapsto \underline{\underline{A}}^0$ be the restriction of f. Note that $B_0 = {}^{1}B_0 \cup {}^{2}B_0$ and the inclusions ${}^{1}B_0 \subset B_0$ and ${}^{2}B_0 \subset B_0$ determine an isomorphism from ${}^{1}f_0 \cup_{{}_0B_0} {}^{2}f_0 : {}^{1}\underline{\underline{B}}_0 \cup_{{}_0B_0} {}^{2}\underline{\underline{B}}_0 \longmapsto \underline{\underline{A}}^0$ to $f_0 : \underline{\underline{B}}_0 \longmapsto \underline{\underline{A}}^0$ over 1_{A^0} (here ${}_{0}B_0 = {}^{1}B_0 \cap {}^{2}B_0 = {}_{0}B \cap B_0$ is a face of both ${}^{1}\underline{\underline{B}}_0$ and ${}^{2}\underline{\underline{B}}_0$, etc.). Consider the diagram

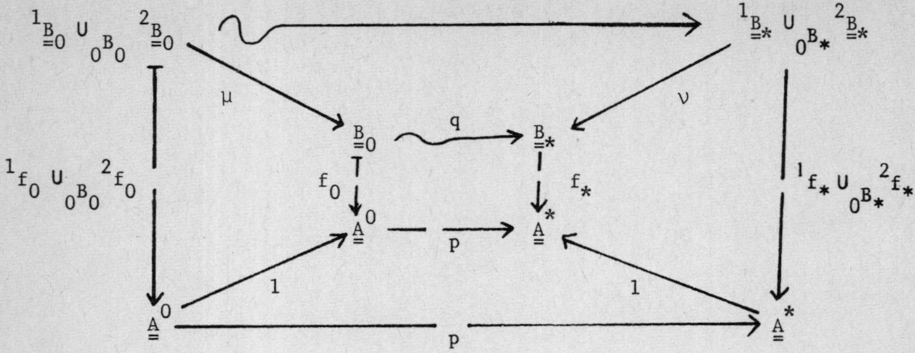

in which the exterior square is $S_1 \cup_{{}_0B_0} S_2$ (see 6.7.4), μ and ν are the weak isomorphisms mentioned above and q is the unique map which makes the diagram commutative. Let S denote the interior square. Since $S_1 \cup_{{}_0B_0} S_2$ is an admissible square (cf. 6.7.4) and (μ, ν, 1, 1) is an isomorphism from $S_1 \cup_{{}_0B_0} S_2$ to S, we can endow S with a canonical structure of an admissible square (see 6.7.2).

Finally, let $\Psi : C(q) \to B_-$ be the unique map which makes commutative the diagram

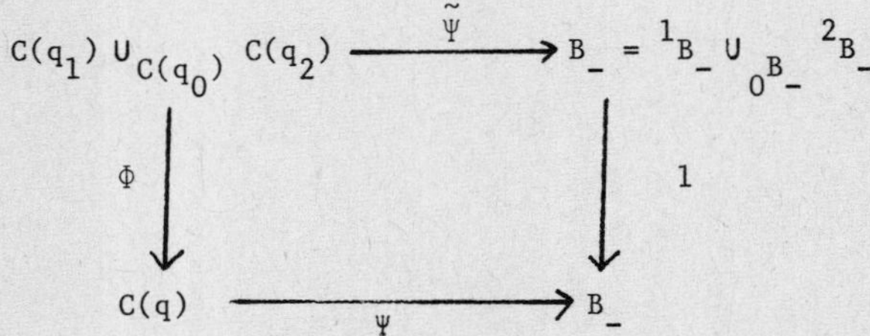

where $q_i : {}^iB_0 \to {}^iB_*$ (i = 1,2) is the quasimorphism of the square S_i, $q_0 : {}_0B_0 \to {}_0B_*$ is the restriction of q_i, $\tilde{\Psi}|C(q_i) = \Psi_i : C(q_i) \to {}^iB_-$ (i = 1,2) and Φ is the homeomorphism defined in 5.3.8. Now one can verify easily that $\nabla_1 \cup_{{}_0B} \nabla_2 = \{\Delta, \underline{\underline{B}}_-, \underline{\underline{B}}_+, S, \Psi\}$ so defined is a decomposition of f.

6.9. THEOREM. Let $f : \underline{\underline{B}} \mapsto \underline{\underline{A}}$ be a proper a.T.m., $\underline{\underline{A}}$ and $\underline{\underline{B}}$ being of finite depth and $B_* \neq \emptyset$. Let $\Delta = \{\underline{\underline{A}}^-, \underline{\underline{A}}^+, p, \varepsilon, \Phi\}$ be a decomposition of $\underline{\underline{A}}$. For any $i \in I_{\underline{\underline{B}}}$ let $f_i : \underline{\underline{B}}_i \longmapsto \underline{\underline{A}}_i$ be the restriction of f (recall that $A_i = A$ if $i \in I^{h,f}$). Let $I \subset I_{\underline{\underline{B}}}$ and assume that for any $i \in I$ the following conditions are satisfied.

(a) $\operatorname{depth}(\underline{\underline{A}}_i) = \operatorname{depth}(\underline{\underline{A}})$;

(b) there is given a decomposition $\nabla_i = \{\Delta_i, \underline{\underline{B}}_i^-, \underline{\underline{B}}_i^+, S_i, \Psi_i\}$ of the a.T.m. $f_i : \underline{\underline{B}}_i \longrightarrow \underline{\underline{A}}_i$, where $\Delta_i = \Delta | A_i$;

(c) if $j \in I$ and $\operatorname{depth}(\underline{\underline{A}}_i | A_i \cap A_j) = \operatorname{depth}(\underline{\underline{A}}_i)$ then $\nabla_i | B_i \cap B_j = \nabla_j | B_i \cap B_j$.

Then there exists a decomposition $\nabla = \{\Delta, \underline{\underline{B}}_-, \underline{\underline{B}}_+, S, \Psi\}$ of f such that $\nabla | B_i = \nabla_i$, $i \in I$.

<u>Proof</u>. If $\operatorname{depth}(\underline{\underline{A}}) = 0$, in view of Remark (1) in 6.8.2, the theorem follows from Theorem 6.5. We assume therefore that $\operatorname{depth}(\underline{\underline{A}}) > 0$.

We shall proceed by induction on $\operatorname{depth}(\underline{\underline{B}}_*)$. The case $\operatorname{depth}(\underline{\underline{B}}_*) = 0$ is "easy" and left to the reader (the arguments are similar to those which follow, but much simpler). Thus we shall assume that $m = \operatorname{depth}(\underline{\underline{B}}_*) > 0$ and that the theorem is true for any proper a.T.m. $f' : \underline{\underline{B}}' \longmapsto \underline{\underline{A}}'$ with $\operatorname{depth}(\underline{\underline{B}}'_*) < m$. In order to point out the main steps of the proof, we shall also assume that $I = \emptyset$ (however, at a certain point, we shall use the inductive hypothesis in its all generality). At the end we shall indicate the necessary changes for handling the general case.

Consider the a.s. $\underline{\underline{B}}_*$ and the weak morphism $f_* = f | B_* : \underline{\underline{B}}_* \longmapsto \underline{\underline{A}}^*$. By Theorem 6.2 there exists an f_*-compatible decomposition $\Delta_* = \{\underline{\underline{B}}_*^-, \underline{\underline{B}}_*^+, p_*, \delta, \Phi_*\}$ of $\underline{\underline{B}}_*$.

For any $Y \in \mathcal{B}$ choose now a continuous $\beta(Y) : Y \to R_+^*$ such that all the control conditions involved hold on $T_Y^{\beta(Y)}$ and

(6.9.1) $$T_Y^{\beta(Y)} \cap B_* \subset U_{\Delta_*}, \quad Y \in \mathcal{B}_*^*,$$

(6.9.2) $$\varphi_{\Delta_*} | T_Y^{\beta(Y)} \cap B_* = (\rho_Y/\delta \circ \pi_Y) | T_Y^{\beta(Y)} \cap B_*, \quad Y \in \mathcal{B}_*^*$$

(6.9.3) $$\pi_{\Delta_*} | T_Y^{\beta(Y)} \cap B_* = \pi_Y | T_Y^{\beta(Y)} \cap B_*, \quad Y \in \mathcal{B}_*^*,$$

(6.9.4) $$2\beta(Y) < \delta|Y, \quad Y \in \mathcal{B}_*^* .$$

Choose also a controlled $\mu : A^* \to R_+^*$ such that, for any $X \in A^*$, all the control conditions involved hold on $T_X^{\mu|X}$, $A^{\mu)} \subset U_\Delta$, $\pi_\Delta|T_X^{\mu|X} = \pi_X|T_X^{\mu|X}$, $\varphi_\Delta|T_X^{\mu|X} = (\rho_X/\varepsilon \circ \pi_X)|T_X^{\mu|X}$ and $f^{-1}(A^{\mu)}) \subset \bigcup_{Z \in \mathcal{B}_*} T_Z^{\beta(Z)}$ (the last condition can be fulfilled because f is proper). In view of Lemma 6.8.10, it is sufficient to construct a decomposition of f of the form $\nabla = \{\Delta_{\mu/2}, \ldots\}$. Therefore, there is no loss of generality in assuming that

(6.9.5) $$U_\Delta = A^{\mu)} = A,$$

(6.9.6) $$\pi_\Delta = \pi^{\mu)} \quad \text{and} \quad \varphi_\Delta = \rho^{\mu)}/\varepsilon \circ \pi^{\mu)} ,$$

(6.9.7) $$B = \bigcup_{Z \in \mathcal{B}_*} T_Z^{\beta(Z)} .$$

In order to simplify the notation, we shall assume that

$$T_Z^{\beta(Z)} = T_Z , \quad Z \in \mathcal{B}$$

and

$$T_X^{\mu|X} = T_X , \quad X \in A^* .$$

Let $Z_1, Z_2 \in \mathcal{B}_* \setminus \mathcal{B}_*^*$, $Z_1 < Z_2$ (then $f(Z_1)$ and $f(Z_2)$ are contained in the same stratum X of $\underline{\underline{A}}^*$). Let $b \in \pi_{Z_1}^{-1}(Z_1 \cap U_{\Delta_*}) \cap \pi_{Z_2}^{-1}(Z_2 \cap U_{\Delta_*})$. Since φ_{Δ_*} is controlled on $U_{\Delta_*} \setminus B_*^*$, we have

$$\varphi_{\Delta_*}(\pi_{Z_2}(b)) = \varphi_{\Delta_*}(\pi_{Z_1}(\pi_{Z_2}(b))) .$$

Since $\pi_{Z_1}(\pi_{Z_2}(b)) = \pi_{Z_1}(b)$, it follows that

$$\varphi_{\Delta_*}(\pi_{Z_2}(b)) = \varphi_{\Delta_*}(\pi_{Z_1}(b)) .$$

Thus if we denote

$$\tilde{W} = \bigcup_{\substack{Z \in \mathcal{B}_* \\ Z \notin \mathcal{B}_*^*}} \pi_Z^{-1}(Z \cap U_{\Delta_*})$$

we can define $\tilde{\varphi}: \tilde{W} \to R_+^*$ by setting

$$\tilde{\varphi}(b) = \varphi_{\Delta_*}(\pi_Z(b)), \quad b \in \pi_Z^{-1}(Z \cap U_{\Delta_*}) .$$

Similarly define $\tilde{\pi} : \tilde{W} \to B_*^*$ by

$$\tilde{\pi}(b) = \pi_{\Delta_*}(\pi_Z(b)), \quad b \in \pi_Z^{-1}(Z \cap U_{\Delta_*}) .$$

Let $Y \in \mathcal{B}_*^*$. One checks immediately that

$$\tilde{\pi}|\tilde{W} \cap T_Y = \pi_Y|\tilde{W} \cap T_Y$$

(use the definition of $\tilde{\pi}$ and (6.9.3)). Similarly

$$\tilde{\varphi}|\tilde{W} \cap T_Y = (\rho_Y/\delta \circ \pi_Y)|\tilde{W} \cap T_Y$$

(use the definition of $\tilde{\varphi}$ and (6.9.2)). Thus if we set $V = \tilde{W} \cup (\bigcup_{Y \in \mathcal{B}_*^*} T_Y)$ we can define $\varphi: V \to R_+$ and $\pi : V \to B_*^*$ by

$$\varphi|\tilde{W} = \tilde{\varphi}, \qquad \varphi|T_Y = \rho_Y/\delta \circ \pi_Y \quad (Y \in \mathcal{B}_*^*) ,$$

$$\pi|\tilde{W} = \tilde{\pi}, \qquad \pi|T_Y = \pi_Y, \quad (Y \in \mathcal{B}_*^*) .$$

It is obvious that π is a weak morphism from $\underline{\underline{B}}|V$ to $\underline{\underline{B}}_*^*$. Also $\varphi|\tilde{W}$ is controlled.

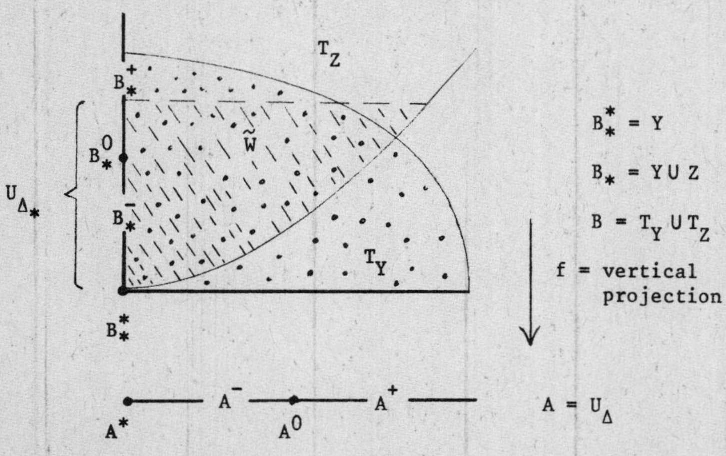

Consider the fibre product $\underline{\underline{P}} = \underline{\underline{B}}_*^* \times_{A^*} \underline{\underline{A}}$ with respect to f_*^* : $B_*^* \to A^*$ and $\pi_\Delta : A = U_\Delta \to A^*$; since $\operatorname{depth}(\underline{\underline{B}}_*) = 0$, $\underline{\underline{P}}$ is an a.s. (see 5.9.3). The decomposition Δ of $\underline{\underline{A}}$ determines a decomposition $\Delta^P = \{\underline{\underline{P}}^-, \underline{\underline{P}}^+, p^P, \varepsilon^P, \Phi^P\}$ of $\underline{\underline{P}}$ as in 6.1.10. Notice that we can take $U_{\Delta^P} = P$ and $\xi_{\Delta^P} = (0 \times \xi_\Delta)|P \smallsetminus P^*$.

Since $f_*^* \circ \pi = \pi_\Delta \circ (f|V)$ (use the fact that f is a weak morphism, the f_*-compatibility of Δ_* and (6.4.3)) we can define $g : V \to P$ by $g(b) = (\pi(b), f(b))$, $b \in V$. Let $Z \in \mathcal{B}_* \smallsetminus \mathcal{B}_*^*$ and $X \in \mathcal{A}^*$ with $f(Z) \subset X$. Consider the commutative diagram

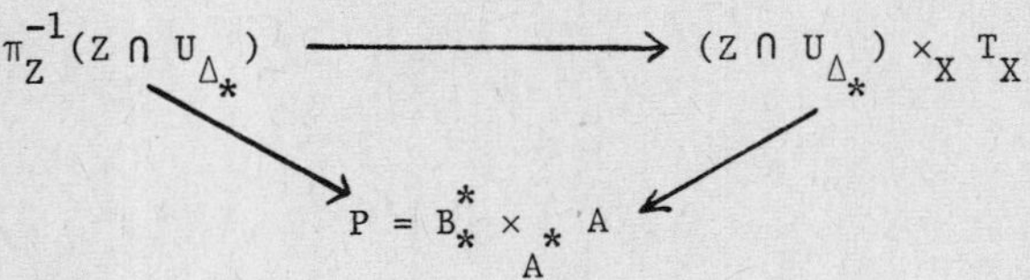

$$\begin{array}{ccc} \pi_Z^{-1}(Z \cap U_{\Delta_*}) & \longrightarrow & (Z \cap U_{\Delta_*}) \times_X T_X \\ & \searrow \quad \swarrow & \\ & P = B_*^* \times_{A^*} A & \end{array}$$

in which

- the fibre product $(Z \cap U_{\Delta_*}) \times_X T_X$ is taken with respect to $f|Z \cap U_{\Delta_*} : Z \cap U_{\Delta_*} \to X$ and π_X ;

- the horizontal map is given by $b \mapsto (\pi_Z(b), f(b))$; in view of 5.9.5 and Remark (2) in 5.5 it is an a.T.m.;

- the right oblique map is given by $(b,a) \longmapsto (\pi(b), a)$; it is easily seen to be an a.T.m.;

- the left oblique map is the restriction of g ; in view of 5.9.9 it is an a.T.m.

Thus $g|\tilde{W}$ is locally an a.T.m. and Remark (1) in 5.5 implies that $g|\tilde{W} : \underline{\underline{B}}|\tilde{W} \longrightarrow \underline{\underline{P}}$ is an a.T.m. Since for any $Y \in \mathcal{B}_*^*$, $g|T_Y$ is an a.T.m. from $\underline{\underline{B}}|T_Y$ to $\underline{\underline{P}}$ (cf. 5.9.5), it follows that g itself is an a.T.m. from $\underline{\underline{V}}$ to $\underline{\underline{P}}$, where $\underline{\underline{V}} = \underline{\underline{B}}|V$.

The decomposition Δ_* of $\underline{\underline{B}}_*$ being f_*-compatible we can assume, possibly after shrinking U_{Δ_*} , that $df \cdot \xi_{\Delta_*} = 0$. Since $d\pi \cdot \xi_{\Delta_*} = d\pi_{\Delta_*} \cdot \xi_{\Delta_*} = 0$, it follows that $dg \cdot \xi_{\Delta_*} = 0$. By Proposition 5.6

applied to $g|\tilde{W}$, there exists $\eta \in \mathcal{X}^{g}_{\underline{\underline{V}}}(\tilde{W})$ such that $dg \cdot \eta = 0$ and $\eta | U_{\Delta_*} \setminus B^*_* = \xi_{\Delta_*}$. We can now shrink U_{Δ_*} and the $\beta(Z)$'s (and also μ) such that the control conditions involved in the definition of η hold on $\tilde{W} \cap T_Z$ ($= \tilde{W} \cap T_Z^{\beta(Z)}$ by our convention) for any $Z \in \mathcal{B}$ with $Z \cap \tilde{W} \neq \emptyset$.

Set $B^0 = \varphi^{-1}(1)$ and $B^- = \varphi^{-1}((-\infty, 1])$; by (6.9.4), $B^0 \subset \tilde{W}$. It is also clear that B^- is closed in V ; as a matter of fact B^- is closed in B too (the verification is tedious but standard; the assumption (6.9.7) is essential).

Let $b \in \pi_Z^{-1}(Z \cap U_{\Delta_*})$, $Z \in \mathcal{B}_* \setminus \mathcal{B}^*_*$. We have

$$d\,\varphi \cdot \eta(b) = d\,\varphi_{\Delta_*} \cdot (d\pi_Z \cdot \eta(b)) = d\,\varphi_{\Delta_*} \cdot \eta(\pi_Z(b)) = d\,\varphi_{\Delta_*} \cdot \xi_{\Delta_*}(\pi_Z(b)) = d/dt .$$

Thus

$$d\,\varphi \cdot \eta = d/dt \quad \text{on} \quad \tilde{W} . \tag{6.9.8}$$

We can therefore apply Proposition 5.11 ; it follows that B^- can be endowed with a w.a.s. structure $\underline{\underline{B}}^-$, $g^- = g|B^-$ is an a.T.m. from $\underline{\underline{B}}^-$ to $\underline{\underline{P}}$, B^0 is a g-horizontal face of $\underline{\underline{B}}^-$ and $g^0 = g|B^0$ is an a.T.m. from $\underline{\underline{B}}^0 = \underline{\underline{B}}^-|B^0 = \underline{\underline{B}}|B^0$ to $\underline{\underline{P}}$.

Since $\mathcal{X}^{g}_{\underline{\underline{V}}}(\tilde{W}) \subset \mathcal{X}^{f}_{\underline{\underline{B}}}(\tilde{W})$ and $dg \cdot \eta = 0$ implies $df \cdot \eta = 0$, we can apply Proposition 5.11 to f too. Let $B^+ = B \setminus (B^- \setminus B^0)$. It follows that we can endow B^+ with a w.a.s. structure $\underline{\underline{B}}^+$ such that $f^+ = f|B^+$ is an a.T.m. from $\underline{\underline{B}}^+$ to $\underline{\underline{A}}$ and $f^- = f|B^-$ is an a.T.m. from $\underline{\underline{B}}^-$ to $\underline{\underline{A}}$ (the w.a.s. structures obtained on B^- by applying Proposition 5.11 to g and f are equal since both of them verify (i) and (iii) of the mentioned proposition). Moreover, B^0 is an f-horizontal face of both $\underline{\underline{B}}^+$ and $\underline{\underline{B}}^-$, $\underline{\underline{B}}^+|B^0 = \underline{\underline{B}}^-|B^0$ and the inclusions $B^- \subset B$ and $B^+ \subset B$ determine an isomorphism from $f^- \cup_{B^0} f^+ : \underline{\underline{B}}^- \cup_{B^0} \underline{\underline{B}}^+ \longmapsto \underline{\underline{A}}$ to f.

Since $V \cap B_* = U_{\Delta_*}$ and $\varphi|U_{\Delta_*} = \varphi_{\Delta_*}$ it follows that $(B^0)_* = (B_*)^0$ and thus $\operatorname{depth}((\underline{\underline{B}}^0)_*) < \operatorname{depth}(\underline{\underline{B}}_*)$. By induction there exists a decomposition $\tilde{\nabla}^0 = \{\Delta^P, \underline{\underline{B}}^0_-, \underline{\underline{B}}^0_+, S^0, \Psi^0\}$ of $g^0 : \underline{\underline{B}}^0 \longmapsto \underline{\underline{P}}$. Set

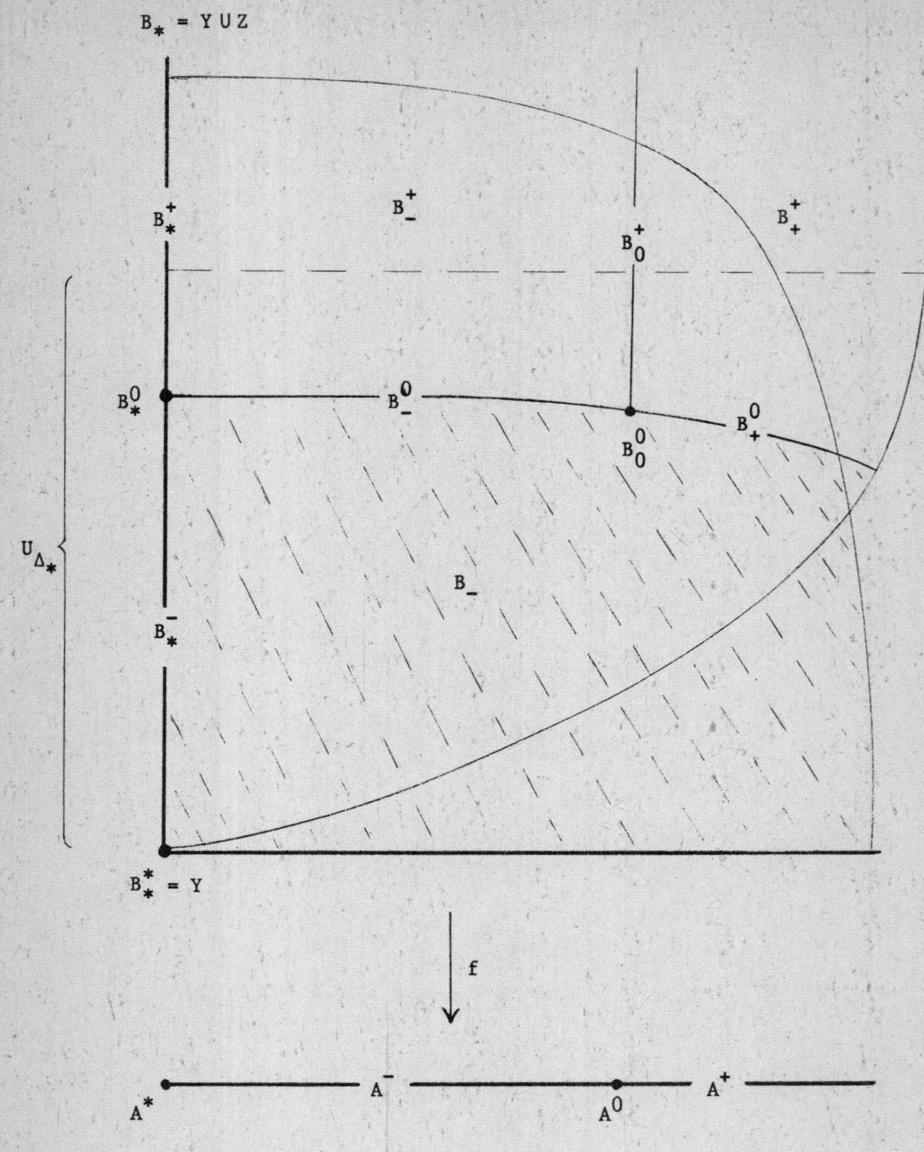

$B_* = Y \cup Z$
B_*^+
B_-^+
B_0^+
B_+^+
B_*^0
B_-^0
B_0^0
B_+^0
U_{Δ_*}
B_-
B_*^-
$B_*^* = Y$
f
A^*
A^-
A^0
A^+

$\nabla^0 = \{\Delta, \underline{\underline{B}}^0_-, \underline{\underline{B}}^0_+, S^0, \Psi^0\}$; clearly ∇^0 is a decomposition of $f^0 : \underline{\underline{B}}^0 \vdash\!\longrightarrow \underline{\underline{A}}$. Since $(\underline{\underline{B}}^+)_* = \underline{\underline{B}}^+_*$ and $\mathrm{depth}(\underline{\underline{B}}^+_*) < \mathrm{depth}(\underline{\underline{B}}_*)$ we can apply the inductive hypothesis to $f^+ : \underline{\underline{B}}^+ \vdash\!\longrightarrow \underline{\underline{A}}$ and find a decomposition $\nabla^+ = \{\Delta, \underline{\underline{B}}^+_-, \underline{\underline{B}}^+_+, S^+, \Psi^+\}$ of f^+ such that $\nabla^+|B^0 = \nabla^0$. Since f and g are proper, we may assume that $U_{\nabla^+} = (f^+)^{-1}(U_\Delta) = B^+$ and $U_{\tilde{\nabla}^0} = g^{-1}(U_{\nabla^P}) = B^0$.

Consider now $\lambda : \tilde{D} \to \tilde{W}$, the flow associated to η. Since $\eta|U_{\Delta_*} \setminus B^*_* = \xi_{\Delta_*}$ it follows that

$$B^0_* \times (-1,0] \subset \tilde{D}.$$

By Lemma 6.10 there exists a continuous $\nu : B^0 \to R^*_+$ such that

$$\nu(b) = 1, \quad b \in B^0_*,$$

$$\nu|B^0 \setminus B^0_* \text{ is controlled},$$

$$B^0 \times (-\nu,0) \subset \tilde{D}.$$

There exists also a controlled $\alpha : B^0 \to R^*_+$ such that

$$B^0 \times (-\nu,\alpha) \subset \tilde{D}.$$

Let $W' = \lambda(B^0 \times (-\nu,\alpha))$ and $\tilde{\lambda} = \lambda|B^0 \times (-\nu,\alpha) : B^0 \times (-\nu,\alpha) \to W'$. A direct verification (by now it is standard) shows that $\tilde{\lambda}$ is a weak isomorphism from $\underline{\underline{B}}^0 \times (-\nu,\alpha)$ to $\underline{\underline{B}}|W'$ (notice that W' is open in B) and $d\tilde{\lambda} \cdot (0 \times d/dt) = \eta$.

Using arguments similar to those in Lemma 6.10 we can construct a continuous $\nu' : B^0 \to R^*_+$ such that

$$\nu'(b) = 1, \quad b \in B^0_*,$$

$$\nu'|B^0 \setminus B^0_* \text{ is controlled},$$

$$\nu'(b) < \nu(b), \quad b \in B^0 \setminus B^0_*.$$

Set $\beta = \nu/\nu'$. One can now construct (the construction is standard) an isomorphism $F : \underline{\underline{B}}^0 \times (-\beta, \alpha) \to \underline{\underline{B}}^0 \times (-\nu, \alpha)$ with the following properties:

(i) $F(b,t) = (b, F_0(b,t))$;

(ii) $F(b,t) = (b,t)$, $(b,t) \in B^0 \times (-\nu', \alpha)$.

Let $\zeta = d(\tilde{\lambda} \circ F) \cdot (0 \times d/dt) \in \mathcal{X}^g_{\underline{\underline{V}}}(W') \subset \mathcal{X}^f_{\underline{\underline{B}}}(W')$. From (ii) above and the definition of $\tilde{\lambda}$,

$$\zeta = \eta \quad \text{near} \quad B^0 \cup (\bar{B}_* \smallsetminus \overset{*}{B}_*) .$$

Notice also that $\beta(b) = 1$ for $b \in B^0_*$, $\beta(b) > 1$ for $b \in B^0 \smallsetminus B^0_*$ and $B^0 \times (-\beta, \alpha) \subset D_\zeta$.

Return now to the decomposition $\tilde{\nabla}_0$ of $g^0 : \underline{\underline{B}}^0 \vdash\!\!\to \underline{\underline{P}}$ and recall that $U_{\tilde{\nabla}^0} = B^0$. Let $\xi_{\tilde{\nabla}^0} \in \mathcal{X}^{g^0}_{\underline{\underline{B}}^0}(B^0 \smallsetminus B^0_*)$ be its first associated vector field. Define $\xi' \in \mathcal{X}^f_{\underline{\underline{B}}}(W')$ by

$$\xi' = d(\tilde{\lambda} \circ F) \cdot (\xi_{\tilde{\nabla}^0} \times 0).$$

From the construction, it is evident that

$$df \cdot \xi' = \xi_\Delta$$

and

$$d\pi \cdot \xi' = 0 .$$

Consider the a.s. $\underline{\underline{P}} = \underline{\underline{B}}^*_* \times_{A^*} \underline{\underline{A}}$ and the vector field $0 \times \xi_\Delta \in \mathcal{X}_{\underline{\underline{B}}_* \times \underline{\underline{A}}}(B^*_* \times (A \smallsetminus A^*))$. Since $d\pi_\Delta \cdot \xi_\Delta = 0$ it follows that

$$(0 \times \xi_\Delta)|P \smallsetminus P^* \in \mathcal{X}_P(P \smallsetminus P^*) .$$

By Proposition 5.6, there exists $\xi'' \in \mathcal{X}^{g^-}_{\underline{\underline{B}}^-}(B^- \smallsetminus B^-_*)$ such that $dg^- \cdot \xi'' = 0 \times \xi_\Delta$, or equivalently

$$df \cdot \xi'' = \xi_\Delta$$

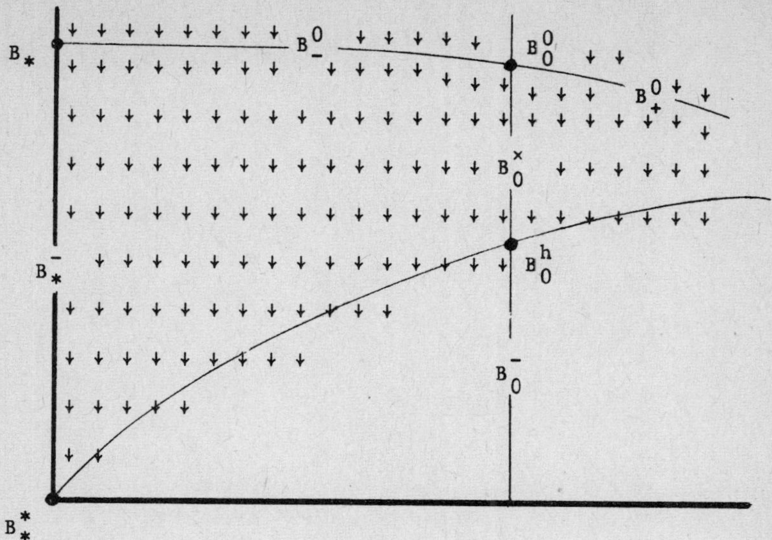

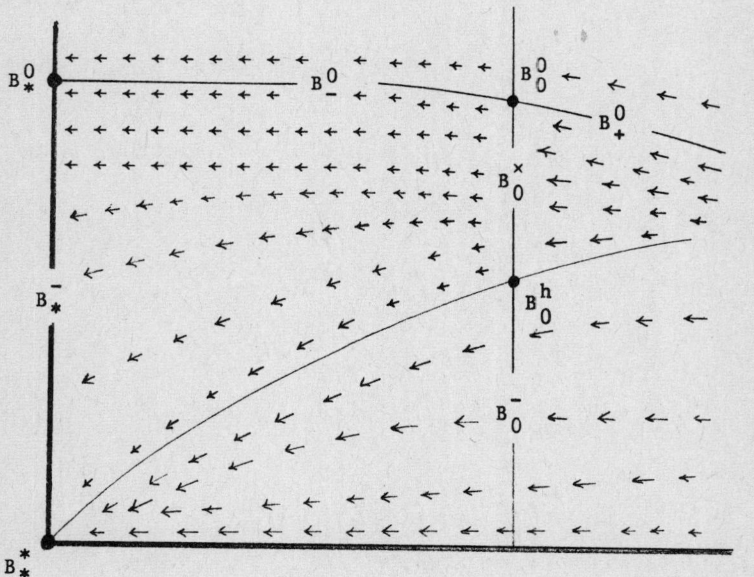

The vector fields ζ and ξ .

and

$$d\pi \cdot \xi'' = 0 .$$

Let $W = (\tilde{\lambda} \circ F)(-\sqrt{\beta}, \alpha)$. Using a partition of unity argument, we can patch together ξ' and ξ'' and obtain $\xi^- \in X_{\underline{\underline{B}}^-}(B^- \smallsetminus B^-_*)$ such that

$$df \cdot \xi^- = \xi_\Delta ,$$

$$d\pi \cdot \xi^- = 0 ,$$

$$\xi^- | W \cap (B^- \smallsetminus B^-_*) = \xi' | W \cap (B^- \smallsetminus B^-_*) .$$

Since $\underline{\underline{B}} = \underline{\underline{B}}^+ \cup_{B^0} \underline{\underline{B}}^-$ and $U_{\nabla^+} = B^+$, we may define $\xi \in X^f_{\underline{\underline{B}}}(B \smallsetminus B_*)$ by $\xi | B^+ \smallsetminus B^* = \xi_{\nabla^+}$ and $\xi | B^- \smallsetminus B_* = \xi^-$ (the definition is correct because $\xi_{\nabla^+} | B^0 \smallsetminus B_* = \xi_{\nabla^0} = \xi_{\tilde{\nabla}^0} = \xi' | B^0 \smallsetminus B_* = \xi^- | B^0 \smallsetminus B_*$ and ξ_{∇^+} and ξ^- are parallel to the faces). Clearly $df \cdot \xi = \xi_\Delta$. By Proposition 5.10, we can endow $B_- = f^{-1}(A^-)$ and $B_+ = f^{-1}(A^+)$ with w.a.s. structures $\underline{\underline{B}}_-$ and $\underline{\underline{B}}_+$ respectively which verify (ii) of the definition of a decomposition of an a.T.m. (see 6.8.2). Moreover, these w.a.s. structures can be chosen such that the vector field associated to the face $B_0 = B_- \cap B_+ = f^{-1}(A^0)$ of $\underline{\underline{B}}_-$ (resp. $\underline{\underline{B}}_+$) is a restriction of ξ.

Define $\psi \in C^\infty_{\underline{\underline{B}}}(W)$ by the relation

$$\psi(\tilde{\lambda}(F(b,t))) = 1+t, \quad (b,t) \in B^0 \times (-\sqrt{\beta}, \alpha) .$$

It is clear that

$$d\psi \cdot \zeta = d/dt$$

$$\psi = \varphi \quad \text{near} \quad B^0 \cup (B^-_* \smallsetminus B^*_*)$$

and

$$\psi^{-1}(1) = \varphi^{-1}(1) = B^0$$

Set now $B^+_0 = B^+ \cap B_0$ $(= (B^+)_0)$, $B^0_0 = B^0 \cap B_0$,

$B_0^\times = B_0 \cap \psi^{-1}([0,1]) = (\tilde{\lambda} \circ F)(B_0^0 \times [0,1])$, $B_0^h = B_0 \cap \psi^{-1}(0)$ and $B_0^- = (B_0 \setminus (B_0^+ \cup B_0^\times)) \cup B_0^h$. Since $df \cdot \eta = 0$, from the construction of ζ it follows that $df \cdot \zeta = 0$ and therefore $\zeta | B_0 \in \mathcal{X}^{f_0}_{\underline{\underline{B}}_0}(W' \cap B_0)$. We can apply Proposition 5.11 (to ψ and ζ) and endow B_0^+, $B_0^\times$ and B_0^- with w.a.s. structures $\underline{\underline{B}}_0^+$, $\underline{\underline{B}}_0^\times$ and $\underline{\underline{B}}_0^-$ respectively which verify condition (3) of the definition of an admissible square (with $f : \underline{\underline{B}} \vdash\!\!\rightarrow \underline{\underline{A}}$ replaced by $f_0 : \underline{\underline{B}}_0 \vdash\!\!\rightarrow \underline{\underline{A}}^0$, see 6.3.1). Moreover we can choose these w.a.s. structures such that the vector field associated to the face B_0^0 (resp. B_0^0, B_0^h, B_0^h) of $\underline{\underline{B}}_0^+$ (resp. $\underline{\underline{B}}_0^\times, \underline{\underline{B}}_0^\times, \underline{\underline{B}}_0^-$) is a restriction of ζ.

Define $\theta : B_0^h \times [0,1] \to B_0^\times$ by $\theta(b,t) = \lambda_\zeta(b,t)$ and notice that θ is a weak isomorphism from $\underline{\underline{B}}_0^h \times [0,1]$ to $\underline{\underline{B}}_0^\times$. We can define $q^\times : B_0^\times \to B_*^-$ by setting

$$q^\times(\theta(b,t)) = \Phi_*([q^+(b),t]),\ (b,t) \in B_0^h \times [0,1]$$

(here $\Phi_* : \underline{\underline{C}}(p_*)^\delta \to \underline{\underline{B}}_*^-$ is the isomorphism given by the decomposition Δ_* of $\underline{\underline{B}}_*$ and $q^+ : \underline{\underline{B}}_0^+ \rightsquigarrow \underline{\underline{B}}_*^+$ is the quasimorphism of the square S^+ of the decomposition ∇^+ of $f^+ : \underline{\underline{B}}^+ \vdash\!\!\rightarrow \underline{\underline{A}}$; $B_0^0 = (q^+)^{-1}(B_*^0))$.

Observe now that $B_0^- \subset B^- = \varphi^{-1}((-\infty,1]) \subset V$ and we can define $q^- : B_0^- \to B_*^*$ to be the restriction of $\pi : V \to B_*^*$. Clearly q^- is a submersive weak morphism from $\underline{\underline{B}}_0^-$ to $\underline{\underline{B}}_*^*$ sending strata onto strata; B_0^h is the unique q^--horizontal face of $\underline{\underline{B}}_0^-$. Moreover, since f is proper and B_0^- is closed in B_0, one checks easily that q is proper. Consider the square

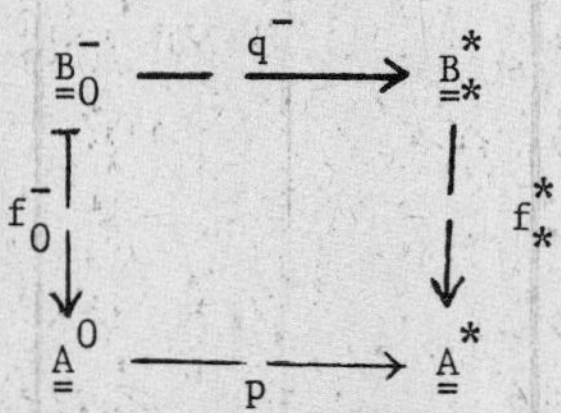

It obviously verifies conditions (i),(iii),(iv) of the definition of a basic square (see 6.7.1). One checks directly that condition (ii) of that definition is also satisfied (or one can use Propositions 5.10 and 5.11 and

the fact that $g : \underline{\underline{V}} \mapsto \underline{\underline{P}}$ is an a.T.m.). Thus the above square is basic.

Define $q : B_0 \to B_*$ by setting

$$q|B_0^+ = q^+, \quad q|B_0^\times = q^\times, \quad q|B_0^- = q^- .$$

Summing up the above remarks it follows that the square

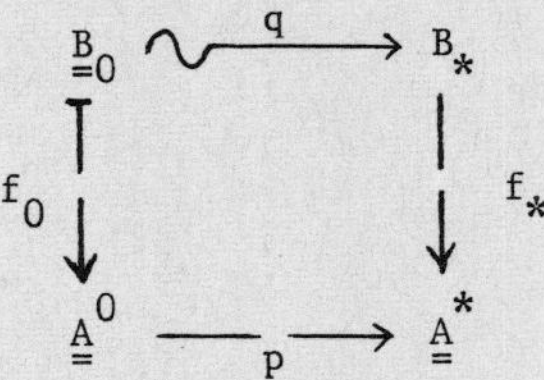

is admissible of depth m.

In order to complete the construction of the decomposition ∇ of f, it remains to construct the weak isomorphism $\Psi : \underline{\underline{C}}(q) \mapsto \underline{\underline{B}}_-$. To this end we set

$$\Psi([b,t]) = \lambda_\xi(b,t-1), \quad b \in B_0, \quad 0 < t \leq 1,$$

$$\Psi([b,0]) = q(b), \quad b \in B_0 ,$$

$$\Psi([b,0]) = b, \quad b \in B_* .$$

Let $b \in B_0$. Let us check that

$$(*) \qquad \lim_{t \searrow 0} \lambda_\xi(b, t-1) = q(b).$$

Case I: $b \in B_0^+$. Then $\lambda_\xi(b, t-1) = \lambda_{\xi_{\Delta^+}}(b,t-1)$ and $(*)$ is true.

Case II: $b \in B_0^\times$. Then $b = \theta(b_0,s)$ for some $b_0 \in B_0^h$ and $0 \leq s \leq 1$. Since $\lambda_\xi(\theta(b_0,s), t-1) = \theta(\lambda_\xi(b_0, t-1), s)$, this case can be reduced to Case I.

Case III: $b \in B_0^-$. Let (s_n) be a sequence in $(-1, 0)$, converging to -1 and such that the sequence $(\lambda_\xi(b, s_n))$ converges to b_* . Set $b_n = \lambda_\xi(b,s_n)$. Since $df \cdot \xi = \xi_\Delta$ it follows that $b_* \in B_*$. Notice that $N = B^+ \cup \lambda_\zeta(B^0 \times (-1,0])$ is an open neighborhood of $B_* \setminus B_*^*$ in B and any $y \in N \smallsetminus B_*$ is of the form $y = \lambda_\xi(y_0, t)$ with $y_0 \in B_0 \smallsetminus B_0^-$ and $t > -1$. Assume that $b_* \in B_* \smallsetminus B_*^*$. Then, for a sufficiently large n, $b_n \in N \smallsetminus B_*$ and therefore $b_n = \lambda_\xi(b_{n,0}, s)$ with $b_{n,0} \in B_0 \smallsetminus B_0^-$. Since on the other hand $b_n = \lambda_\xi(b, s_n)$ with $b \in B_0^-$, we get a contradiction. It follows that $b_* \in B_*^*$. Since $\xi | B^- \smallsetminus B_* = \xi^-$ and $d\pi \cdot \xi^- = 0$, we deduce that

$$\pi(b_n) = \pi(\lambda_\xi(b, s_n)) = \pi(b) = q(b)$$

and therefore

$$b_* = \pi(b_*) = \lim \pi(b_n) = q(b).$$

Now an easy topological argument (based on the fact that f is proper) shows that this implies (*).

There are no more difficulties in verifying that Ψ is a weak isomorphism and that $\nabla = \{\Delta, \underline{\underline{B}}_-, \underline{\underline{B}}_+, S, \Psi\}$ is a decomposition of f.

Return now to the general case when $I \neq \emptyset$. The proof follows exactly the same lines as above. In order that $\nabla | B_i = \nabla_i$ for any $i \in I$, the data constructed above must satisfy the following additional conditions:

(a) $\Delta_* | (B_i)_* = \Delta_{S_i}$ for any $i \in J = \{j \in I \, ; \, (B_i)_* = B_i \cap B_*\} = \{j \in I \, ; \, \mathrm{depth}(\underline{\underline{B}}_i)_* = \mathrm{depth}(\underline{\underline{B}}_*)\}$ (Δ_{S_i} is the decomposition of $(\underline{\underline{B}}_i)_*$ involved in the definition of the admissible square S_i);

(b) $(B_i)_-^\times \smallsetminus B_*^* \subset W$, $\psi = \psi_{\nabla_i}$ near $(B_i)_-^\times \smallsetminus B_*^*$, $\zeta = \zeta_{\Delta_i}$ near $(B_i)_-^\times \smallsetminus B_*^*$, $\xi | B_i \smallsetminus B_*^* = \xi_{\nabla_i}$ $(i \in J)$;

(c) $(B_i)^+ = B_i \cap B^+$ and $\nabla^+ | (B_i)^+ = (\nabla_i)^+$, for any $i \in I$ (see 6.8.8 for the meanings of $(B_i)^+$ and $(\nabla_i)^+$).

Next some remarks about the possibility of satisfying these conditions:

- (a) follows from Theorem 6.2;

- the first condition in (c) raises no problem; the second one follows from the inductive hypothesis;

- for (b), one has to "thicken" W_{∇_i} near B_i, then to extend in the obvious way ψ_{∇_i}, ζ_{∇_i} and ξ_{∇_i} on this thickening of W_{∇_i} and finally to patch them to ψ, ζ and ξ respectively, which are defined as above. Q.E.D.

LEMMA 6.10. Let $\underline{\underline{A}}$ be a w.a.s., $A_0 \subset A$ be a closed subset and $D \subset A \times R_+$ be an open subset such that $A \times \{0\} \subset D$ and $(A_0 \times R_+) \cap D = A_0 \times [0,1)$. Then there exists a continuous $f : A \to R_+^*$ such that

(1) $f|A \smallsetminus A_0 \in C^\infty_{\underline{\underline{A}}}(A \smallsetminus A_0)$;

(2) $f(a) = 1$, $a \in A_0$;

(3) $A \times [0,f) \subset D$.

Proof. For any $a \in A$ set $t_a = \sup\{t \in R_+ ; \{a\} \times [0,t) \subset D\}$. Let $U_n = \{a \in A; t_a > 1 - 1/n\}$, $n \geq 2$; then U_n is open in A, $A_0 \subset U_n$ and $A = \cup U_n$. Choose open subsets V_n $(n \geq 1)$ of A such that $V_1 = A$, $cl_A(V_n) \subset U_n \cap V_{n-1}$ $(n \geq 2)$ and $A_0 = \cap V_n$. Let also $\varphi_n \in C^\infty_{\underline{\underline{A}}}(A)$ verify

(i) $\varphi_n(a) = 1$, $a \in V_n$;

(ii) $\varphi_n(a) = 0$, $a \notin U_n \cap V_{n-1}$

(iii) $\varphi_n(a) \in [0,1]$, $a \in A$.

Choose $f_1 \in C^\infty_{\underline{\underline{A}}}(A)$ such that $f_1(a) > 0$ for any $a \in A$ and $A \times [0,f_1) \subset D$. For $n \geq 2$ define inductively $f_n \in C^\infty_{\underline{\underline{A}}}(A)$ by

$$f_n = (1-1/n)\varphi_n + (1-\varphi_n)f_{n-1} .$$

It is obvious that $f_n(a) > 0$ for any $a \in A$, $f_n(a) = 1 - 1/n$ for any $a \in V_n$ and $A \times [0,f_n) \subset D$. Finally, let $f : A \to R_+^*$ be given by $f(a) = \lim_{n\to\infty} f_n(a)$. A straightforward verification shows that f has the required properties. Q.E.D.

6.11. As before $f : \underline{\underline{B}} \vdash\!\rightarrow \underline{\underline{A}}$ is an a.T.m., $\underline{\underline{B}}$ and $\underline{\underline{A}}$ being of finite depth. Let $n = \text{depth}(\underline{\underline{A}}) \geq 0$.

6.11.1. A <u>total decomposition</u> of f is a sequence $\mathcal{D} = \{\nabla^1,\ldots,\nabla^{n+1}\}$, where $\nabla^1 = \{\Delta^1, \underline{\underline{B}}_-, \underline{\underline{B}}_+, S, \Psi\}$ is a decomposition of f and, inductively if $n > 0$, $\mathcal{D}_+ = \{\nabla^2,\ldots,\nabla^{n+1}\}$ is a total decomposition of $f_+ = f|B_+ : \underline{\underline{B}}_+ \vdash\!\rightarrow \underline{\underline{A}}^+$ (since $\text{depth}(\underline{\underline{A}}^+) = n-1$, the definition makes sense by induction). If $n = 0$, then $\mathcal{D} = \{\nabla^1\}$, $B_+ = \emptyset = A^+$, $\underline{\underline{B}}_- = \underline{\underline{B}}$, $\underline{\underline{A}}_- = \underline{\underline{A}}$ and $\mathcal{D}_S$ is a regular and f-compatible total decomposition of $\underline{\underline{B}}$ (see 6.7.1 (7)); thus in this case a total decomposition of f reduces to a regular and f-compatible total decomposition of $\underline{\underline{B}}$.

A total decomposition $\mathcal{D}$ of f determines a total decomposition $\mathcal{D}_{\mathcal{D}}$ of $\underline{\underline{A}}$ as follows: if $n=0$, set $\mathcal{D}_{\mathcal{D}} = \{\Delta^1\}$; if $n > 0$, set $\mathcal{D}_{\mathcal{D}} = \{\Delta^1, \Delta^2,\ldots, \Delta^{n+1}\}$, where $\{\Delta^2,\ldots, \Delta^{n+1}\} = \mathcal{D}_{\mathcal{D}_+}$.

6.11.2. Let $\mathcal{D} = \{\nabla^1,\ldots,\nabla^{n+1}\}$ be a total decomposition of f, B_i be a face of $\underline{\underline{B}}$ and $f_i : \underline{\underline{B}}_i \vdash\!\rightarrow \underline{\underline{A}}_i$ be the restriction of f. Let $\text{depth}(\underline{\underline{A}}_i) = n-k$. One defines a total decomposition $\mathcal{D}|B_i = \{\tilde{\nabla}^{k+1},\ldots,\tilde{\nabla}^{n+1}\}$ of f_i as follows. Let $\underline{\underline{A}}^j$ have the same meaning as in 6.3.2 and set $B^j = f^{-1}(A^j)$ and $f^j = f|B^j : B^j \rightarrow A^j$. Then B^j has a canonical w.a.s. structure $\underline{\underline{B}}^j$, f^j is an a.T.m. from $\underline{\underline{B}}^j$ to $\underline{\underline{A}}^j$, $\nabla^j = \{\Delta^j, (\underline{\underline{B}}^j)_-, (\underline{\underline{B}}^j)_+, S^j, \Psi^j\}$ is a decomposition of f^j and $\underline{\underline{B}}^{j+1} = (B^j)_+$ if $j \leq n$. By depth arguments, A_i is a face of $\underline{\underline{A}}^{k+1}$ and therefore B_i is a face of $\underline{\underline{B}}^{k+1}$. Take $\tilde{\nabla}^{k+j} = \nabla^{k+j}|B_i \cap B^{k+j}$, $j = 1,\ldots,n-k+1$. Notice that $\mathcal{D}_{\mathcal{D}|B_i} = \mathcal{D}_{\mathcal{D}}|A_i$.

6.11.3. Two total decompositions $\mathcal{D} = \{\nabla^1,\ldots,\nabla^{n+1}\}$ and $\mathcal{D}' = \{'\nabla^1,\ldots, '\nabla^{n+1}\}$ of f are called <u>equal</u>, denoted $\mathcal{D} = \mathcal{D}'$, if $\nabla^1 = '\nabla^1,\ldots, \nabla^{n+1} = '\nabla^{n+1}$.

6.11.4. Assume now that $\underline{\underline{B}} = '\underline{\underline{B}} \cup_C {''\underline{\underline{B}}}$ and $f = 'f \cup_C {''f}$, where $C = 'B \cap {''B}$ is an $'f$ (resp. $''f$)-horizontal face of $'\underline{\underline{B}}$ (resp. $''\underline{\underline{B}}$) and $'f : '\underline{\underline{B}} \vdash\!\rightarrow \underline{\underline{A}}$ and $''f : {''\underline{\underline{B}}} \vdash\!\rightarrow \underline{\underline{A}}$. Let $'\mathcal{D} = \{'\nabla^1,\ldots,'\nabla^{n+1}\}$ and $''\mathcal{D} = \{''\nabla^1,\ldots, ''\nabla^{n+1}\}$ be total decompositions of $'f$ and $''f$ respectively such that $'\mathcal{D}|C = {''\mathcal{D}}|C$. We shall define inductively (on $\text{depth}(\underline{\underline{A}})$) a total decomposition $'\mathcal{D} \cup_C {}''\mathcal{D} = \{\nabla^1,\ldots,\nabla^{n+1}\}$ of f as follows:

(a) $\nabla^1 = {}'\nabla^1 \cup_C {}''\nabla^1$ (see 6.8.11);

(b) if $n > 0$, then $\underline{\underline{B}}_+ = {}'\underline{\underline{B}}_+ \cup_{C_+} {}''\underline{\underline{B}}_+$, where $C_+ = {}'B_+ \cap C = {}''B_+ \cap C$ is a face of both $'\underline{\underline{B}}_+$ and $''\underline{\underline{B}}_+$; by hypothesis (and by definition) $'\mathcal{D}_+|C_+ = {}''\mathcal{D}_+|C_+$ and thus, by induction, we can consider $'\mathcal{D}_+ \cup_{C_+} {}''\mathcal{D}_+$; then $'\mathcal{D} \cup_C {}''\mathcal{D}$ is determined by (a) and the relation

$$('\mathcal{D} \cup_C {}''\mathcal{D})_+ = {}'\mathcal{D}_+ \cup_{C_+} {}''\mathcal{D}_+ .$$

6.11.5. Let $\underline{\underline{M}}$ be a compact a.s. of depth zero and define $f_M : \underline{\underline{B}} \times \underline{\underline{M}} \longrightarrow \underline{\underline{A}}$ by $f_M(b,x) = f(b)$. Clearly f_M is a proper a.T.m. Given a total decomposition $\mathcal{D} = \{\nabla^1,\ldots,\nabla^{n+1}\}$ of f we define a total decomposition $\mathcal{D}_M = \{\nabla^1_M,\ldots,\nabla^{n+1}_M\}$ of f_M as follows. If $\nabla^1 = \{\Delta^1, \underline{\underline{B}}_-, \underline{\underline{B}}_+, S, \Psi\}$, set $\nabla^1_M = \{\Delta^1, \underline{\underline{B}}_- \times \underline{\underline{M}}, \underline{\underline{B}}_+ \times \underline{\underline{M}}, S_M, \Psi_M\}$, where S_M is the square

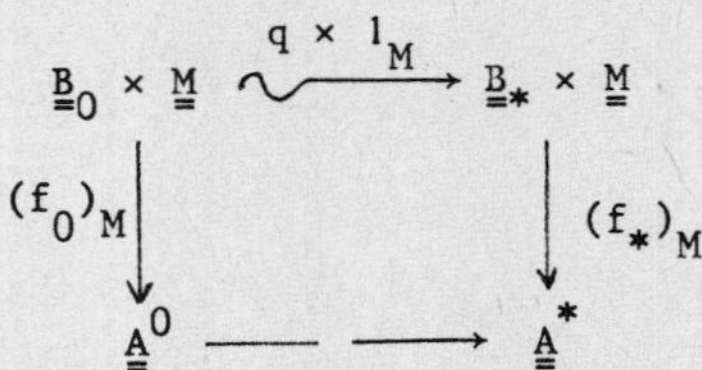

and $\Psi_M : C(q \times 1_M) \to B_- \times M$ is the map given by $\Psi_M([(b,x),t]) = (\Psi([b,t]),x)$. It is obvious that ∇^1_M is a decomposition of f_M (the structure of admissible square on S_M is the obvious one; for example $(S_M)_+ = (S_+)_M$). If $n = 0$, we are done. Otherwise $\mathcal{D}_M$ is determined by ∇^1_M defined above and the relation

$$(\mathcal{D}_M)_+ = (\mathcal{D}_+)_M .$$

6.11.6. Let $f' : \underline{\underline{B}}' \vdash\!\to \underline{\underline{A}}'$ be another a.T.m., let (G,F) be an isomorphism f to f' and let $\mathcal{D}$ be a total decomposition of f. Then there exists an obvious total decomposition $G_*(\mathcal{D})$ of f' (if $\mathcal{D} = \{\nabla^1,\ldots,\nabla^{n+1}\}$ then $G_*(\mathcal{D}) = \{'\nabla^1,\ldots,{}'\nabla^{n+1}\}$ with $'\nabla^1 = G_*(\nabla^1)$, etc.).

6.11.7. Consider the regular square

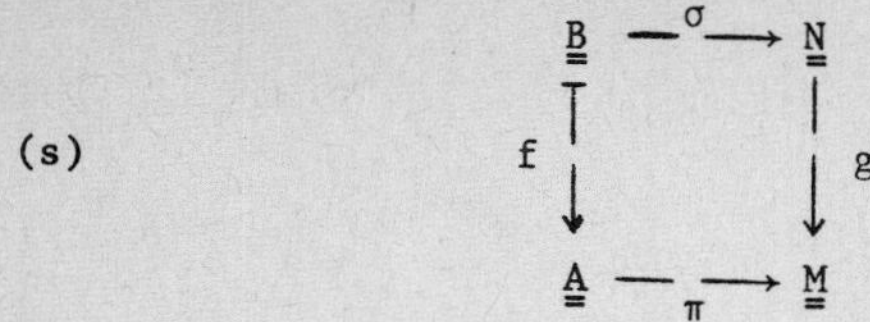

and let $f^s : \underline{\underline{B}} \vdash\to \underline{\underline{A}}^s = \underline{\underline{A}} \times_M \underline{\underline{N}}$ be the a.T.m. given by $f^s(b) = (f(b), \sigma(b))$ (see the Remark following the definition of a regular square). A decomposition $\nabla = \{\Delta, \underline{\underline{B}}_-, \underline{\underline{B}}_+, S, \Psi\}$ of f is called s-<u>compatible</u> if

(i) Δ is π-compatible;

(ii) the total decomposition $\mathcal{D}_S$ of $\underline{\underline{B}}_*$ (see 6.7.1 (7)) is $(\sigma|B_*)$-compatible;

(iii) $d\sigma \cdot \xi_\nabla = 0$ near B_- .

From (iii) it follows that $(\sigma|B_*) \circ \pi_\nabla = \sigma$ near B_- and from (ii) we deduce that

(iv) $d\sigma \cdot \zeta_\nabla = 0$ near $\bar{B}_- \cup \overset{\times}{B}_-$.

Assume that ∇ is s-compatible. Then by (i) we can consider the decomposition $\Delta^s = \Delta^g$ of $\underline{\underline{A}}^s$ (see 6.1.10). A direct verification (using induction on $\operatorname{depth}(\underline{\underline{B}}_*)$) shows that the square

$$(S^s) \qquad \begin{array}{ccc} \underline{\underline{B}}_0 & \overset{q}{\rightsquigarrow} & \underline{\underline{B}}_* \\ (f^s)_0 \downarrow & & \downarrow (f^s)_* \\ (\underline{\underline{A}}^s)^0 & \underset{p^s}{\longrightarrow} & (\underline{\underline{A}}^s)^* \end{array}$$

is admissible (the notation is obvious). It follows that $\nabla^s = \{\Delta^s, \underline{\underline{B}}_-, \underline{\underline{B}}_+, S^s, \Psi\}$ is a decomposition of f^s . From (i) and (ii) we also deduce that $\sigma_+ = \sigma|B_+ : B_+ \to N$ and $\pi^+ = \pi|A^+ : A^+ \to M$ are weak morphisms and the square

$$(s_+) \qquad \begin{array}{ccc} \underline{\underline{B}}_+ & \overset{\sigma_+}{\longrightarrow} & \underline{\underline{N}} \\ f^+ \downarrow & & \downarrow g \\ \underline{\underline{A}} & \underset{\pi^+}{\longrightarrow} & \underline{\underline{M}} \end{array}$$

is regular.

A total decomposition $\mathcal{D} = \{\nabla = \nabla^1, \nabla^2, \ldots, \nabla^{n+1}\}$ of f is called s-<u>compatible</u> if ∇ is s-compatible and, inductively if $n > 0$, the total decomposition $\mathcal{D}_+ = \{\nabla^2, \ldots, \nabla^{n+1}\}$ of f_+ is (s_+)-compatible. If $\mathcal{D}$ is s-compatible then $\mathcal{D}_{\mathcal{D}}$ is π-compatible and we can define a total decomposition $\mathcal{D}^s = \{\tilde{\nabla}^1, \ldots, \tilde{\nabla}^{n+1}\}$ of f^s as follows: $\tilde{\nabla}^1 = \nabla^s$ and, inductively if $n > 0$, $(\mathcal{D}^s)_+ = (\mathcal{D}_+)^{s_+}$. Notice that $\mathcal{D}_{\mathcal{D}^s} = (\mathcal{D}_{\mathcal{D}})^g$.

6.11.8. Let now

$$(s) \qquad \begin{array}{ccc} \underline{\underline{B}} & \overset{\sigma}{\rightsquigarrow} & \underline{\underline{N}} \\ f \downarrow & & \downarrow g \\ \underline{\underline{A}} & \underset{\pi}{\longrightarrow} & \underline{\underline{M}} \end{array}$$

be an admissible square of depth m and, if $m > 0$, let

$$(s^+) \qquad \begin{array}{ccc} \underline{\underline{B}}^+ & \overset{\sigma^+}{\rightsquigarrow} & \underline{\underline{N}}^+ \\ f^+ \downarrow & & \downarrow g^+ \\ \underline{\underline{A}} & \underset{\pi}{\longrightarrow} & \underline{\underline{M}} \end{array}$$

$$(s^\times) \qquad \begin{array}{ccc} \underline{\underline{B}}^\times & \overset{\sigma^\times}{\longrightarrow} & \underline{\underline{N}}^- \\ f^\times \downarrow & & \downarrow g^- \\ \underline{\underline{A}} & \underset{\pi}{\longrightarrow} & \underline{\underline{M}} \end{array}$$

and

$$(s^-) \qquad \begin{array}{ccc} \underline{\underline{B}}^- & \overset{\sigma^-}{\longrightarrow} & \underline{\underline{N}}^* \\ f^- \downarrow & & \downarrow g^* \\ \underline{\underline{A}} & \underset{\pi}{\longrightarrow} & \underline{\underline{M}} \end{array}$$

be the associated squares (s^+ is admissible of depth $m-1$ and s^- is basic, hence regular). A total decomposition $\mathcal{D}$ of f is called s-<u>compatible</u> if either $m = 0$ and $\mathcal{D}$ is s-compatible (in this case s is regular) or $m > 0$ and there exist total decompositions $\mathcal{D}^+$, $\mathcal{D}^\times$ and $\mathcal{D}^-$ of f^+, $f^\times$ and f^- respectively such that

(1) $\mathcal{D}^-$ is s^--compatible (s^- is regular !);

(2) $\mathcal{D}^+$ is s^+-compatible (this makes sense by induction);

(3) $\mathcal{D}^\times = (\theta_s)_*((\mathcal{D}^-|B^h)_{[0,1]})$ (see 6.7.5 and 6.7.6);

(4) $\mathcal{D}^\times|B^0 = \mathcal{D}^+|B^0$;

(5) $\mathcal{D} = \mathcal{D}^+ \cup_{B^0} \mathcal{D}^\times \cup_{B^h} \mathcal{D}^-$.

6.11.9. Let $\mathcal{D} = \{\nabla^1, \ldots, \nabla^{n+1}\}$ be a total decomposition of $f : \underline{\underline{B}} \vdash\!\!\rightarrow \underline{\underline{A}}$ and let $\nabla^1 = \{\Delta^1, \underline{\underline{B}}_-, \underline{\underline{B}}_+, S, \Psi\}$. $\mathcal{D}$ is called <u>regular</u> if either $n = 0$ or $n > 0$ and

(i) $\mathcal{D}$ is a regular total decomposition of $\underline{\underline{A}}$;

(ii) $\mathcal{D}_+$ is a regular total decomposition of $f_+ : \underline{\underline{B}}_+ \vdash\!\!\rightarrow \underline{\underline{A}}^+$ (this makes sense by induction on n);

(iii) $\mathcal{D}_0 = \mathcal{D}_+|B_0$ is an S-compatible total decomposition of $f_0 : \underline{\underline{B}}_0 \vdash\!\!\rightarrow \underline{\underline{A}}^0$ and the total decompositions $\mathcal{D}_0^+$ of $f_0^+ : \underline{\underline{B}}_0^+ \vdash\!\!\rightarrow \underline{\underline{A}}^0$, $\mathcal{D}_0^\times$ of $f_0^\times : \underline{\underline{B}}_0^\times \vdash\!\!\rightarrow \underline{\underline{A}}^0$ and $\mathcal{D}^-$ of $f_0^- : \underline{\underline{B}}_0^- \vdash\!\!\rightarrow \underline{\underline{A}}^0$ associated with $\mathcal{D}_0$ (see 6.11.8) are regular.

6.11.10. Let $\mathcal{D}$ be a regular total decomposition of f, let B_i be a face of $\underline{\underline{B}}$ and let $f_i : \underline{\underline{B}}_i \vdash\!\!\rightarrow \underline{\underline{A}}_i$ be the restriction of f . Then the total decomposition $\mathcal{D}|B_i$ of f_i is regular.

6.11.11. Consider a regular square

$$(s) \qquad \begin{array}{ccc} \underline{\underline{B}} & \xrightarrow{\ \sigma\ } & \underline{\underline{N}} \\ f \big\downarrow & & \big\downarrow g \\ \underline{\underline{A}} & \xrightarrow[\ \pi\ ]{} & \underline{\underline{M}} \end{array}$$

and let $\mathcal{D}$ be a regular and s-compatible total decomposition of f . A direct verification shows that the total decomposition $\mathcal{D}^s$ of $f^s : \underline{\underline{B}} \vdash\!\!\rightarrow \underline{\underline{A}}^s$ (see 6.11.7) is regular.

6.11.12. The notation being as in 6.11.4, assume that the total decompositions $'\mathcal{D}$ and $''\mathcal{D}$ are regular. Then the total decomposition $'\mathcal{D} \cup_C {}''\mathcal{D}$ is also regular.

6.11.13. Consider the submersive weak morphisms $\pi : \underline{\underline{A}} \dashrightarrow \underline{\underline{M}}$ and $g : \underline{\underline{N}} \dashrightarrow \underline{\underline{M}}$, $\underline{\underline{N}}$ and $\underline{\underline{M}}$ being of depth zero and all the faces of $\underline{\underline{A}}$ being π-vertical. Let $\underline{\underline{B}} = \underline{\underline{A}} \times_M \underline{\underline{N}}$ and let $f : \underline{\underline{B}} \dashrightarrow \underline{\underline{A}}$ and $\sigma : \underline{\underline{B}} \dashrightarrow \underline{\underline{N}}$ be the canonical projections. The square

$$(s) \qquad \begin{array}{ccc} \underline{\underline{B}} & \xrightarrow{\sigma} & \underline{\underline{N}} \\ f \downarrow & & \downarrow g \\ \underline{\underline{A}} & \xrightarrow[\pi]{} & \underline{\underline{M}} \end{array}$$

is clearly admissible of depth zero. Given a π-compatible, regular, total decomposition $\mathcal{D}$ of $\underline{\underline{A}}$, there exists a unique s-compatible regular total decomposition $\mathfrak{D} = \mathfrak{D}^{\mathcal{D}}$ of f such that $\mathcal{D}_{\mathfrak{D}} = \mathcal{D}$.

To construct $\mathfrak{D}$ we proceed as follows. Let $\mathcal{D} = \{\Delta^1, \Delta^2, \dots, \Delta^{n+1}\}$. Define $\nabla^1 = \{\Delta^1, \underline{\underline{B}}_-, \underline{\underline{B}}_+, S, \Psi\}$ by setting $\underline{\underline{B}}_- = \underline{\underline{A}}^- \times_M \underline{\underline{N}}$ and $\underline{\underline{B}}_+ = \underline{\underline{A}}^+ \times_M \underline{\underline{N}}$. Then $\underline{\underline{B}}_0 = \underline{\underline{A}}^0 \times_M \underline{\underline{N}}$, $\underline{\underline{B}}_* = \underline{\underline{A}}^* \times_M \underline{\underline{N}}$ and S is the square

$$\begin{array}{ccc} \underline{\underline{B}}_0 = \underline{\underline{A}}^0 \times_M \underline{\underline{N}} & \xrightarrow{\quad q \quad} & \underline{\underline{A}}^* \times_M \underline{\underline{N}} = \underline{\underline{B}}_* \\ f_0 \downarrow & & \downarrow f_* \\ A^0 & \xrightarrow[p]{\quad\quad} & A^* \end{array}$$

the mappings being the obvious ones. Ψ is the obvious homeomorphism of $C(q)$ on B^-.

By induction on n, we can define now $\mathfrak{D}_+ = \mathfrak{D}^{\mathcal{D}^+} = \{\nabla^2, \dots, \nabla^{n+1}\}$. Finally set $\mathfrak{D} = \{\nabla^1, \nabla^2, \dots, \nabla^{n+1}\}$.

6.12. THEOREM. Let $f : \underline{\underline{B}} \vdash\!\rightarrow \underline{\underline{A}}$ be proper, let

$$\text{(s)} \qquad \begin{array}{ccc} \underline{\underline{B}} & \xrightarrow{\ \sigma\ } & \underline{\underline{N}} \\ f \downarrow & & \downarrow g \\ \underline{\underline{A}} & \xrightarrow[\ \pi\ ]{} & \underline{\underline{M}} \end{array}$$

be an admissible square of depth m and let $\mathcal{D}$ be a regular and π-compatible total decomposition of $\underline{\underline{A}}$. Let $I \subset I_{\underline{\underline{B}}}$ and for any $i \in I$ let $\mathfrak{D}_i$ be a regular and $(s|B_i)$-compatible total decomposition of $f_i : \underline{\underline{B}}_i \vdash\!\rightarrow \underline{\underline{A}}_i$. Assume that $\mathcal{D}_{\mathfrak{D}_i} = \mathcal{D}|A_i$ and $\mathfrak{D}_i|B_i \cap B_j = \mathfrak{D}_j|B_i \cap B_j$ for any $i,j \in I$. Then there exists a regular and s-compatible total decomposition $\mathfrak{D}$ of f such that $\mathcal{D}_{\mathfrak{D}} = \mathcal{D}$ and $\mathfrak{D}|B_i = \mathfrak{D}_i$ for any $i \in I$.

<u>Proof</u>. Let $n = \text{depth}(\underline{\underline{A}})$, $\mathcal{D} = \{\Delta^1, \ldots, \Delta^{n+1}\}$ and $\mathfrak{D}_i = \{\nabla_i^1, \ldots, \nabla_i^{n_i+1}\}$, where $n_i = \text{depth}(\underline{\underline{A}}_i) \leq n$. We begin with the following remark.

Assume $m = \text{depth}(\underline{\underline{N}}) = 0$. Then s is regular and we can consider the a.T.m. $f^s : \underline{\underline{B}} \vdash\!\rightarrow \underline{\underline{A}}^s$, the regular total decomposition $\mathcal{D}^g$ of $\underline{\underline{A}}^s$ and the regular total decompositions $\mathfrak{D}_i^{s|B_i}$ of $(f^s)_i : \underline{\underline{B}}_i \vdash\!\rightarrow (\underline{\underline{A}}^s)_i$, $i \in I$. Let $\tilde{\mathfrak{D}}$ be a regular total decomposition of f^s such that $\mathcal{D}_{\tilde{\mathfrak{D}}} = \mathcal{D}^g$ and $\tilde{\mathfrak{D}}|B_i = \mathfrak{D}_i^{s|B_i}$, $i \in I$. Then there exists a unique regular and s-compatible total decomposition $\mathfrak{D}$ of f such that $\tilde{\mathfrak{D}} = \mathfrak{D}^s$, $\mathcal{D}_{\mathfrak{D}} = \mathcal{D}$ and $\mathfrak{D}|B_i = \mathfrak{D}_i$, $i \in I$. Now, using induction on m (as in Steps II, III and IV below), we can see that the Theorem follows from the following weaker assertion :

(*) "Let $f : \underline{\underline{B}} \vdash\!\rightarrow \underline{\underline{A}}$ be proper, let $\mathcal{D}$ be a regular total decomposition of $\underline{\underline{A}}$, let $I \subset I_{\underline{\underline{B}}}$ and for any $i \in I$ let $\mathfrak{D}_i$ be a regular total decomposition of $f_i : \underline{\underline{B}}_i \vdash\!\rightarrow \underline{\underline{A}}_i$. Assume that $\mathcal{D}_{\mathfrak{D}_i} = \mathcal{D}|A_i$ and $\mathfrak{D}_i|B_i \cap B_j = \mathfrak{D}_j|B_i \cap B_j$ for any $i,j \in I$. Then there exists a regular total decomposition $\mathfrak{D}$ of f such that $\mathcal{D}_{\mathfrak{D}} = \mathcal{D}$ and $\mathfrak{D}|B_i = \mathfrak{D}_i$ for any $i \in I$".

We shall prove (*) by induction on $n = \text{depth}(\underline{\underline{A}})$. If $n = 0$, $\underline{\underline{B}}$ is an a.s. and a regular total decomposition of f reduces to a regular and f-compatible total decomposition of $\underline{\underline{B}}$; the assertion follows from Theorem 6.5. Assume now that $n > 0$ and that (*) (and hence the Theorem too) is true for any a.T.m. $f' : \underline{\underline{B}}' \vdash\!\!\rightarrow \underline{\underline{A}}'$ with $\mathrm{depth}(\underline{\underline{A}}') < n$.

Step I. By Theorem 6.5 there exists a decomposition $\nabla^1 = \{\Delta^1, \underline{\underline{B}}_-, \underline{\underline{B}}_+, S, \Psi\}$ of f such that $\nabla^1|B_i = \nabla^1_i$ for any $i \in I$ with $\mathrm{depth}(\underline{\underline{A}}_i) = \mathrm{depth}(\underline{\underline{A}})$.

Step II. Let

$$(S^-) \qquad \begin{array}{ccc} \underline{\underline{B}}_0^- & \xrightarrow{\ q^-\ } & \underline{\underline{B}}_*^* \\ f_0^- \downarrow & & \downarrow f_*^* \\ \underline{\underline{A}}^0 & \xrightarrow[\ p\ ]{} & \underline{\underline{A}}^* \end{array}$$

be the basic square associated to S . Then, by induction, there exists a regular and S^--compatible total decomposition $\mathcal{D}_0^-$ of f_0^- such that $\mathcal{D}_{\mathcal{D}_0^-} = \mathcal{D}^+|A^0$, $\mathcal{D}_0^-|B_i \cap B_0^- = (\mathcal{D}_i)_0^-$, $i \in I$.

Step III. $\mathcal{D}_0^\times = (\theta_S)_*((\mathcal{D}_0^-|B_0^h)_{[0,1]})$ is a regular and $S^\times$-compatible total decomposition of $f_0^\times$ such that $\mathcal{D}_{\mathcal{D}_0^\times} = \mathcal{D}^+|A^0$, $\mathcal{D}_0^\times|B_0^h = \mathcal{D}_0^-|B_0^h$ and $\mathcal{D}_0^\times|B_0^\times \cap B_i = (\mathcal{D}_i)_0^\times$, $i \in I$.

Step IV. By induction there exists a regular and S^+-compatible total decomposition $\mathcal{D}_0^+$ of $f_0^+ : \underline{\underline{B}}_0^+ \vdash\!\!\rightarrow \underline{\underline{A}}^0$ such that $\mathcal{D}_{\mathcal{D}_0^+} = \mathcal{D}^+|A^0$, $\mathcal{D}_0^+|B_0^0 = \mathcal{D}_0^\times|B_0^0$ and $\mathcal{D}_0^+|B_0^+ \cap B_i = (\mathcal{D}_i)_0^+$, $i \in I$.

Step V. $\mathcal{D}_0 = \mathcal{D}_0^+ \cup_{B_0^0} \mathcal{D}_0^\times \cup_{B_0^h} \mathcal{D}_0^-$ is a regular and S-compatible total decomposition of $f_0 : \underline{\underline{B}}_0 \vdash\!\!\rightarrow \underline{\underline{A}}^0$ such that $\mathcal{D}_{\mathcal{D}_0} = \mathcal{D}^+|A^0$ and $\mathcal{D}_0|B_0 \cap B_i = (\mathcal{D}_i)_0|B_0 \cap B_i$.

Step VI. By induction there exists a regular total decomposition $\mathcal{D}_+ = \{\nabla^2, \ldots, \nabla^{n+1}\}$ of $f_+ : \underline{\underline{B}}_+ \vdash\!\!\rightarrow \underline{\underline{A}}^+$ such that $\mathcal{D}_{\mathcal{D}_+} = \mathcal{D}^+$, $\mathcal{D}_+|B_0 = \mathcal{D}_0$ and $\mathcal{D}_+|B_+ \cap B_i = (\mathcal{D}_i)_+|B_+ \cap B_i$, $i \in I$.

Step VII. $\mathcal{D} = \{\nabla^1, \nabla^2, \ldots, \nabla^{n+1}\}$ is a regular total decomposition of f such that $\mathcal{D}_{\mathcal{D}} = \mathcal{D}$ and $\mathcal{D}|B_i = \mathcal{D}_i$, $i \in I$. Q.E.D.

6.13. Let $\mathcal{D} = \{\nabla^1, \ldots, \nabla^{n+1}\}$ be a total decomposition of f. Let $\nabla^1 = \{\Delta^1, \underline{\underline{B}}_-, \underline{\underline{B}}_+, S, \Psi\}$ and let $\mathcal{D}_S$ be the corresponding total decomposition of $\underline{\underline{B}}_*$. If $n = 0$ set $c(\underline{\underline{B}},\mathcal{D}) = c(\underline{\underline{B}},\mathcal{D}_S)$; if $n > 0$ define inductively $c(\underline{\underline{B}},\mathcal{D}) = c(\underline{\underline{B}}_*,\mathcal{D}_S) \sqcup c(\underline{\underline{B}}_+,\mathcal{D}_+)$. It is obvious from the definition that $c(\underline{\underline{B}},\mathcal{D})$ is a manifold with faces and that $c(f,\mathcal{D}) = f|c(\underline{\underline{B}},\mathcal{D}) : c(\underline{\underline{B}},\mathcal{D}) \longrightarrow c(\underline{\underline{A}},\mathcal{D}_{\mathcal{D}})$ is a submersion compatible with the faces. $c(\underline{\underline{B}},\mathcal{D})$ is called the $\mathcal{D}$-<u>core</u> of $\underline{\underline{B}}$.

If B_i is a face of $\underline{\underline{B}}$ then it is easily seen that $B_i \cap c(\underline{\underline{B}},\mathcal{D}) = c(\underline{\underline{B}}_i,\mathcal{D}|B_i)$ and $c(\underline{\underline{B}}_i,\mathcal{D}|B_i)$ is a face of $c(\underline{\underline{B}},\mathcal{D})$.

Assume next that $\underline{\underline{B}} = {}'\underline{\underline{B}} \cup_C {}''\underline{\underline{B}}$ and $f = {}'f \cup_C {}''f$, where ${}'f : {}'\underline{\underline{B}} \vdash\!\!\to \underline{\underline{A}}$ and ${}''f : {}''\underline{\underline{B}} \vdash\!\!\to \underline{\underline{A}}$ are a.T.m.'s. Let ${}'\mathcal{D}$ and ${}''\mathcal{D}$ be total decompositions of ${}'f$ and ${}''f$ respectively such that ${}'\mathcal{D}|C = {}''\mathcal{D}|C$ and let $\mathcal{D} = {}'\mathcal{D} \cup_C {}''\mathcal{D}$ (see 6.11.4). Then $c(\underline{\underline{C}},{}'\mathcal{D}|C) = c(\underline{\underline{C}},{}''\mathcal{D}|C)$ and $c(\underline{\underline{B}},\mathcal{D}) = c({}'\underline{\underline{B}},{}'\mathcal{D}) \cup_{c(\underline{\underline{C}},{}'\mathcal{D}|C)} c({}''\underline{\underline{B}},{}''\mathcal{D})$ (this follows directly from the definitions).

6.14. Let now

$$(s)\qquad \begin{array}{ccc} \underline{\underline{B}} & \overset{\sigma}{\leadsto\!\!\longrightarrow} & \underline{\underline{N}} \\ {\scriptstyle f}\big\downarrow & & \big\downarrow{\scriptstyle g} \\ \underline{\underline{A}} & \underset{\pi}{\longrightarrow} & \underline{\underline{M}} \end{array}$$

be **an** admissible square of depth m and let $\mathcal{D}$ be an s-compatible total decomposition of f ; let also (s^+), $(s^\times)$, (s^-), $\mathcal{D}^+$ and $\mathcal{D}^-$ be the associated data (see 6.11.8). If $m = 0$, set $c_s(\underline{\underline{B}},\mathcal{D}) = c(\underline{\underline{B}},\mathcal{D})$ and if $m > 0$, define inductively $c_s(\underline{\underline{B}},\mathcal{D}) = c(\underline{\underline{B}}^-,\mathcal{D}^-) \sqcup c_{s^+}(\underline{\underline{B}}^+,\mathcal{D}^+)$. $c_s(\underline{\underline{B}},\mathcal{D})$ is called the $(s,\mathcal{D})$-<u>core</u> of $\underline{\underline{B}}$; it is clearly a manifold with faces. If $m > 0$, notice that

$$c(\underline{\underline{B}},\mathcal{D}) = c(\underline{\underline{B}}^-,\mathcal{D}^-) \cup_{c(\underline{\underline{B}}^h,\mathcal{D}^-|B^h)} c(\underline{\underline{B}}^\times,\mathcal{D}^\times) \cup_{c(\underline{\underline{B}}^0,\mathcal{D}^+|B^0)} c(\underline{\underline{B}}^+,\mathcal{D}^+)$$

and $c(\underline{\underline{B}}^\times,\mathcal{D}^\times)$ is diffeomorphic through θ_s to $c(\underline{\underline{B}}^h,\mathcal{D}^-|B^h) \times [0,1]$. By taking the restrictions of f, g, σ and π we obtain the regular square of manifolds with faces

$$(s_{\mathcal{D}}) \qquad \begin{array}{ccc} c_s(\underline{\underline{B}},\mathcal{D}) & \xrightarrow{c_s(\sigma,\mathcal{D})} & c(\underline{\underline{N}},\mathcal{D}_s) \\ {\scriptstyle c_s(f,\mathcal{D})}\downarrow & & \downarrow{\scriptstyle c(g,\mathcal{D}_s)} \\ c(\underline{\underline{A}},\mathcal{D}_{\mathcal{D}}) & \xrightarrow[c(\pi,\mathcal{D}_{\mathcal{D}})]{} & M \end{array}$$

If B_i is a face of $\underline{\underline{B}}$, we can consider $s|B_i$ and $\mathcal{D}|B_i$. It is obvious that

$$c_{s|B_i}(\underline{\underline{B}}_i,\mathcal{D}|B_i) = c_s(\underline{\underline{B}},\mathcal{D}) \cap B_i$$

and this manifold with faces is a face of $c_s(\underline{\underline{B}},\mathcal{D})$.

6.15. The assumptions and notation are as in 6.11.13. Then

$$c(\underline{\underline{B}},\mathcal{D}) = c_s(\underline{\underline{B}},\mathcal{D}) = c(\underline{\underline{A}},\mathcal{D}) \times_M \underline{\underline{N}} .$$

7. TRIANGULATION OF ABSTRACT STRATIFICATIONS

In this chapter we shall prove that any a.s. of finite depth can be triangulated. All notions concerning simplicial complexes and triangulations of topological spaces can be found in the Appendix.

7.1.1. A <u>relative manifold</u> (with corners) is a pair of topological spaces $(V,\delta V)$ such that δV is a closed subset of V and $V \setminus \delta V$ is a manifold with corners.

Examples. (1) If X is a manifold with corners, then $(X,\emptyset)$ is a relative manifold.

(2) Let $\underline{\underline{A}}$ be a w.a.s. and X be a stratum of $\underline{\underline{A}}$. Set $\overline{X} = \mathrm{cl}_A(X)$ and $\overline{X} = \overline{X} \setminus \delta\overline{X}$. Then $(\overline{X}, \delta\overline{X})$ is a relative manifold.

7.1.2. Let $(V,\delta V)$ be a relative manifold. A triangulation (K,φ) of V is called <u>smooth</u> if K contains a subcomplex δK such that

(i) $\varphi(|\delta K|) = \delta V$;

(ii) for any closed simplex $\sigma < K$ the restriction of φ to $|\sigma| \setminus |\delta K|$ is smooth (see [Mu], Section 8.1);

(iii) for any $x \in |K| \setminus |\delta K|$ the differential $d\varphi_x$ of φ at x is injective (see [Mu], Section 8.2).

7.1.3. If X is a manifold with corners, a smooth triangulation of X is just a smooth triangulation of the relative manifold $(X,\emptyset)$.

7.1.4. Let now X be a manifold with faces. A triangulation (K,φ) of X is called compatible with the faces if for any face X_i of X there exists a subcomplex K_i of K such that

(1) $\varphi(|K_i|) = X_i$;

(2) the map $x \mapsto \varphi^{-1}(p_{X_i}(\varphi(x)))$ from $\varphi^{-1}(U_{X_i}) \subset |K|$ to $|K_i|$ is piecewise linear near $|K_i|$;

(3) the map $x \mapsto r_{X_i}(\varphi(x))$ from $\varphi^{-1}(U_{X_i})$ to R is piecewise linear near $|K_i|$.

A triangulation (K,φ) of X is called good if it is smooth and compatible with the faces. Any subdivision of a good triangulation is still a good triangulation. If Y is another manifold with faces and $f : X \to Y$ is smooth and compatible with the faces, a triangulation $\{(K,\varphi),(L,\psi)\}$ of f is called good if both (K,φ) and (L,ψ) are good.

7.1.5. Let (K,φ) be a good triangulation of the manifold with faces X and let X_i be a face of X . Then X_i has a natural structure of a manifold with faces (see Remark (2) in 4.1) and the triangulation $(K_i,\varphi\big||K_i|)$ of X_i (cf. (1) in 7.1.4) is obviously good. We shall denote $(K_i,\varphi\big||K_i|)$ by $(K,\varphi)|X_i$.

7.2.1. From now on $\underline{\underline{A}}$ will denote an a.s. of finite depth, $\underline{\underline{M}}$ an a.s. of depth zero and $f : \underline{\underline{A}} \dashrightarrow \underline{\underline{M}}$ a proper and submersive weak morphism. We shall fix a regular total decomposition $\mathcal{D} = \{\Delta^1, \Delta^2,\ldots, \Delta^{n+1}\}$ of $\underline{\underline{A}}$ with $\Delta^1 = \{\underline{\underline{A}}^-,\underline{\underline{A}}^+,p,\varepsilon,\Phi\}$, $\mathcal{D}^+ = \{\Delta^2,\ldots, \Delta^{n+1}\}$ and $\mathcal{D}^0 = \mathcal{D}^+|A^0$. In order to simplify the notation we set $c(\underline{\underline{A}}) = c(\underline{\underline{A}},\mathcal{D})$, $c(\underline{\underline{A}}^0) = c(\underline{\underline{A}}^0,\mathcal{D}^0)$ and $c(\underline{\underline{A}}^+) = c(\underline{\underline{A}}^+,\mathcal{D}^+)$. If A_i is a face of $\underline{\underline{A}}$, set $\mathcal{D}_i = \mathcal{D}|A_i$ and

$c(\underline{\underline{A}}_i) = c(\underline{\underline{A}}_i, \mathcal{D}_i)$. If $\mathcal{D}$ is f-compatible, we shall consider the following restrictions of f : $f^*: A^* \to M$, $f^0 : A^0 \longrightarrow M$, $f^- : A^- \longrightarrow M$, $f^+ : A^+ \longrightarrow M$, $c(f) : c(\underline{\underline{A}}) \longrightarrow M$, $c(f^0) : c(\underline{\underline{A}}^0) \longrightarrow M$, $c(f^+) : c(\underline{\underline{A}}^+) \longrightarrow M$, $f_i : A_i \longrightarrow M_i$ and $c(f_i) : c(\underline{\underline{A}}_i) \longrightarrow M_i$. Let also $c(p) : c(\underline{\underline{A}}^0) \longrightarrow A^*$ be the restriction of p.

7.2.2. Let (K,φ) be a good triangulation of $c(\underline{\underline{A}})$. Since $c(\underline{\underline{A}})$ is the disjoint sum of A^* and $c(\underline{\underline{A}}^+)$ we can consider the restrictions (K^+,φ^+) and (K^*,φ^*) of (K,φ) to $c(\underline{\underline{A}}^+)$ and A^* respectively. If $\mathcal{D}$ is f-compatible and $\{(K,\varphi),(L,\psi)\}$ is a good triangulation of $c(f)$ it is obvious that $\{(K^+,\varphi^+),(L,\psi)\}$ and $\{(K^*,\varphi^*),(L,\psi)\}$ are good triangulations of $c(f^+)$ and f^* respectively.

7.2.3. Let (K,φ) be a good triangulation of $c(\underline{\underline{A}})$ and let A_i be a face of $\underline{\underline{A}}$. As noticed in 6.6.3, $c(\underline{\underline{A}}_i)$ is a face of $c(\underline{\underline{A}})$. By 7.1.5, $(K_i,\varphi_i) = (K,\varphi)|c(\underline{\underline{A}}_i)$ is a good triangulation of $c(\underline{\underline{A}}_i)$. Assume now that $\mathcal{D}$ is f-compatible and let $\{(K,\varphi),(L,\psi)\}$ be a good triangulation of $c(f)$. Let $(L_i,\psi_i) = (L,\psi)|M_i$ be the restriction of (L,ψ) to M_i. Clearly $\{(K_i,\varphi_i),(L_i,\psi_i)\}$ is a good triangulation of $c(f_i)$; we shall denote it $\{(K,\varphi),(L,\psi)\}|c(f_i)$.

7.2.4. A good triangulation (K,φ) of $c(\underline{\underline{A}})$ is called <u>regular</u> if either $\mathrm{depth}(\underline{\underline{A}}) = 0$ or $\mathrm{depth}(\underline{\underline{A}}) > 0$ and

(i) $(K^+,\varphi^+) = (K,\varphi)|c(\underline{\underline{A}}^+)$ is a regular triangulation of $c(\underline{\underline{A}}^+)$;

(ii) if $(K^0,\varphi^0) = (K^+,\varphi^+)|c(\underline{\underline{A}}^0)$, then $\{(K^0,\varphi^0),(K^*,\varphi^*)\}$ is a triangulation of $c(p)$.

7.2.5. Let (K,φ) be a regular triangulation of $c(\underline{\underline{A}})$ and let A_i be a face of $\underline{\underline{A}}$. A direct argument shows that $(K,\varphi)|c(\underline{\underline{A}}_i)$ is a regular triangulation of $c(\underline{\underline{A}}_i)$.

7.3. LEMMA. Assume $\mathcal{D}$ is f-compatible. Let $\{(K,\varphi),(L,\psi)\}$ be a good triangulation of $c(f)$ and let (L',ψ) be a subdivision of (L,ψ). Let

$I \subset I_{\underline{\underline{A}}}$ and for any $i \in I$ let (K_i, φ_i) be a subdivision of $(K,\varphi)|c(\underline{\underline{A}}_i)$ such that

(a) $\{(K_i,\varphi_i),(L',\psi)|M_i\}$ is a good triangulation of $c(f_i)$;

(b) $(K_i,\varphi_i)|c(\underline{\underline{A}}) \cap A_i \cap A_j = (K_j,\varphi_j)|c(\underline{\underline{A}}) \cap A_i \cap A_j$ for any $i, j \in I$ (the restrictions make sense !).

Then there exists a subdivision (K',φ) of (K,φ) such that

$\{(K',\varphi),(L',\psi)\}$ is a good triangulation of $c(f)$ and $(K',\varphi)|c(\underline{\underline{A}}_i) = (K_i,\varphi_i)$, $i \in I$. If (L',ψ) and (K_i,φ_i), $i \in I$, are baricentric subdivisions then (K',φ) may be chosen to be a baricentric subdivision of (K,φ) .

The <u>proof</u> of this lemma uses the same type of induction as that used in the proof of Theorem 6.5. Since it raises no difficulty, we omit it.

7.4. LEMMA. Let N be a manifold with faces, let $g : N \to M$ be a proper submersion compatible with the faces and let (L,ψ) be a good triangulation of M. Let also $I \subset I_N$ and for any $i \in I$ let (K_i,φ_i) be a good triangulation of N_i such that

(a) $\{(K_i,\varphi_i),(L,\psi)|M_i\}$ is a good triangulation of $g_i : N_i \to M_i$, the restriction of g ;

(b) $(K_i,\varphi_i)|N_i \cap N_j = (K_j,\varphi_j)|N_i \cap N_j$, $j \in I$.

Then there exists a subdivision (L',ψ) of (L,ψ) and a good triangulation (K,φ) of N such that

(1) $\{(K,\varphi),(L',\psi)\}$ is a good triangulation of g ;

(2) $(K,\varphi)|N_i$ is a subdivision of (K_i,φ_i), $i \in I$.

<u>Proof</u>. One argues exactly as in the proof of Corollary 2.3 of [P]. I want to point out the following two facts:

(i) As in [P] one has to construct "product type" triangulations of U_{N_i} , $i \in I$, and then fit them together to obtain a triangulation of a neighborhood U of $\bigcup_{i \in I} N_i$. This is possible because of condition (b) above and of conditions (2) and (3) in 7.1.4. Notice also that, if U is sufficiently small, then the restriction of f to U will be piecewise linear with respect to the so constructed triangulation of U and (L,ψ).

(ii) In [P] one obtains a triangulation of N with respect to which f is only piecewise linear. To obtain our stronger assertion, one has to apply Theorem 3.6 in [Hu] . Q.E.D.

Remark. The above Lemma also follows from $[J_1]$.

7.5. LEMMA. Assume $\mathcal{D}$ is f-compatible and let (L,ψ) be a good triangulation of M . Let also $I \subset I_{\underline{\underline{A}}}$ and for any $i \in I$ let (K_i,φ_i) be a regular triangulation of $c(\underline{\underline{A}}_i)$ such that $\{(K_i,\varphi_i),(L,\psi)|M_i\}$ is a triangulation of $c(f_i)$ and, for any $j \in I$, the restrictions of (K_i,φ_i) and (K_j,φ_j) to $c(\underline{\underline{A}}) \cap A_i \cap A_j$ are equal. Then there exists a subdivision (L',ψ) of (L,ψ) and a regular triangulation (K,φ) of $c(\underline{\underline{A}})$ such that $\{(K,\varphi),(L',\psi)\}$ is a triangulation of $c(f)$ and for any $i \in I$ $(K,\varphi)|c(\underline{\underline{A}}_i)$ is a subdivision of (K_i,φ_i) .

Proof. We proceed by induction on $\text{depth}(\underline{\underline{A}})$. If $\text{depth}(\underline{\underline{A}}) = 0$, the assertion follows from Lemma 7.4. Assume now that $\text{depth}(\underline{\underline{A}}) > 0$.

Step I. By Lemma 7.4 we can find a subdivision (L^1,ψ) of (L,ψ) and a good triangulation $(\hat{K}^*,\varphi^*)$ of A^* such that f^* is simplicial with respect to $(\hat{K}^*,\varphi^*)$ and (L^1,ψ), and $(\hat{K}^*,\varphi^*)|A^* \cap A_i$ is a subdivision of $(K_i,\varphi_i)|A^* \cap A_i$ if $A^* \cap A_i \neq \emptyset$, $i \in I$.

Step II. Let $i \in I$ be such that $A^* \cap A_i \neq \emptyset$. Then $A^0 \cap A_i \neq \emptyset$ and is a face of both $\underline{\underline{A}}^0$ and $\underline{\underline{A}}_i$; clearly $c(\underline{\underline{A}}^0) \cap A_i = c(\underline{\underline{A}}_i) \cap A^0 = c(\underline{\underline{A}}^0|A^0 \cap A_i, \mathcal{D}^0|A^0 \cap A_i)$. Since the triangulation (K_i,φ_i) of $c(\underline{\underline{A}}_i)$ is regular, it follows that $c(p_i) = p|c(\underline{\underline{A}}^0) \cap A_i : c(\underline{\underline{A}}^0) \cap A_i \to A^* \cap A_i$ is simplicial with respect to the corresponding restrictions (which exist) of (K_i,φ_i) . Now, as in Lemma 7.3, we can subdivide and obtain a regular triangulation (K_i^0,φ_i^0) of $c(\underline{\underline{A}}^0) \cap A_i$ such that $\{(K_i^0,\varphi_i^0),(\hat{K}^*,\varphi^*)|A^* \cap A_i\}$ is a triangulation of $c(p_i)$ and, for any $j \in I$ with $A^0 \cap A_i \cap A_j \neq \emptyset$, the restrictions of (K_i^0,φ_i^0) and (K_j^0,φ_j^0) to $c(\underline{\underline{A}}^0) \cap A_i \cap A_j$ are equal.

Step III. By induction there exist a subdivision $(\tilde{K}^*,\varphi^*)$ of $(\hat{K}^*,\varphi^*)$ and a regular triangulation (K^0,φ^0) of $c(\underline{\underline{A}}^0)$ such that $\{(K^0,\varphi^0),(\tilde{K}^*,\varphi^*)\}$ is a triangulation of $c(p)$ and $(K^0,\varphi^0)|c(\underline{\underline{A}}^0) \cap A_i$ is a subdivision of (K_i^0,φ_i^0) if $i \in I$ and $A^* \cap A_i \neq \emptyset$. There is no loss of generality in assuming that $\{(K^0,\varphi^0),(L^1,\psi)\}$ is a triangulation of $c(f^0) : c(\underline{\underline{A}}^0) \to M$. Indeed, if this not so, we can first subdivide L^1 and $\tilde{K}^*$ to make f^* simplicial (Theorem 3.6 in [Hu]) and then subdivide K^0 to make $c(p)$ simplicial (Lemma 7.3); since $f|A^0 = (f|A^*)\circ p$, the assertion follows.

Step IV. Arguing again as in Lemma 7.3 we can find regular triangulations (K_i^+,φ_i^+) of $c(\underline{\underline{A}}_i^+) = c(\underline{\underline{A}}_i^+,\mathcal{D}^+|A_i^+)$, where $\underline{\underline{A}}_i^+ = \underline{\underline{A}}^+|A^+ \cap A_i$, such that

(1) (K_i^+,φ_i^+) is a subdivision of $(K_i,\varphi_i)|c(\underline{\underline{A}}_i^+)$;

(2) $\{(K_i^+,\varphi_i^+),(L^1,\psi)|M_i\}$ is a triangulation of $c(f_i^+) : c(\underline{\underline{A}}_i^+) \to M_i$;

(3) if $i, j \in I$ and $c(\underline{\underline{A}}^+) \cap A_i \cap A_j \neq \emptyset$, then the restrictions of (K_i^+,φ_i^+) and (K_j^+,φ_j^+) to $c(\underline{\underline{A}}^+) \cap A_i \cap A_j$ are equal (note that $c(\underline{\underline{A}}^+) \cap A_i \cap A_j$ is a face of both $c(\underline{\underline{A}}_i^+)$ and $c(\underline{\underline{A}}_j^+)$ and therefore the above restrictions make sense);

(4) if $i \in I$ and $A^* \cap A_i \neq \emptyset$ (this is equivalent to $A^0 \cap A_i \neq$ $\neq \emptyset$), then the restrictions of (K_i^+,φ_i^+) and (K^0,φ^0) to $c(\underline{\underline{A}}^0) \cap A_i =$ $= c(\underline{\underline{A}}_i^+) \cap A^0$ are equal.

Step V. By induction there exist a subdivision (L',ψ) of (L^1,ψ) and a regular triangulation $(\tilde{K}^+,\varphi^+)$ of $c(\underline{\underline{A}}^+)$ such that $\{(\tilde{K}^+,\varphi^+),(L',\psi)\}$ is a triangulation of $c(f^+)$, $(\tilde{K}^+,\varphi^+)|c(\underline{\underline{A}}^0)$ is a subdivision of (K^0,φ^0) and, for any $i \in I$, $(\tilde{K}^+,\varphi^+)|c(\underline{\underline{A}}_i^+)$ is a subdivision of (K_i^+,φ_i^+).

Step VI. By Lemma 7.3 we can subdivide $(\tilde{K}^*,\varphi^*)$ and obtain a a good triangulation (K^*,φ^*) of A^* such that f^* is simplicial with respect to (K^*,φ^*) and (L',ψ). Next we can subdivide $(\tilde{K}^+,\varphi^+)|c(\underline{\underline{A}}^0)$

to make $c(p)$ simplicial and then extend this subdivision to a subdivision (K^+,φ^+) of $c(\underline{\underline{A}}^+)$ which is regular and such that $\{(K^+,\varphi^+),(L',\psi)\}$ is a triangulation of $c(f^+)$. Finally, since $c(\underline{\underline{A}})$ is the disjoint sum of A^* and $c(\underline{\underline{A}}^+)$, (K^*,φ^*) and (K^+,φ^+) fit together and give rise to a regular triangulation (K,φ) of $c(\underline{\underline{A}})$ with the required properties.

Q.E.D.

7.6. A <u>smooth triangulation</u> of a w.a.s. $\underline{\underline{B}}$ is by definition a triangulation (K,φ) of B verifying: for any stratum X of $\underline{\underline{B}}$ there exists a subcomplex K_X of K such that $(K_X,\varphi||K_X|)$ is a smooth triangulation of the relative manifold $(\overline{X},\delta\overline{X})$ (see Example (2) in 7.1.1).

7.7. PROPOSITION. Let (K,φ) be a regular triangulation of $c(\underline{\underline{A}})$. Then there exists a canonical (but not unique) construction of a smooth triangulation $(\hat{K},\hat{\varphi})$ of $\underline{\underline{A}}$, called a <u>canonical extension</u> of (K,φ), such that

(1) $(\hat{K},\hat{\varphi})|c(\underline{\underline{A}})$ exists and is a subdivision of (K,φ);

(2) if A_i is a face of $\underline{\underline{A}}$ and (K_i,φ_i) is the regular triangulation $(K,\varphi)|c(\underline{\underline{A}}_i)$ of $c(\underline{\underline{A}}_i)$ (see 7.2.5), then $(\hat{K},\hat{\varphi})|A_i$ exists and is a canonical extension of (K_i,φ_i);

(3) if $\mathcal{D}$ is f-compatible and $\{(K,\varphi),(L,\psi)\}$ is a good triangulation of $c(f)$, then f is piecewise linear with respect to $(\hat{K},\hat{\varphi})$ and (L,ψ) (as a matter of fact we can construct $(\hat{K},\hat{\varphi})$ such that f is simplicial with respect to $(\hat{K},\hat{\varphi})$ and a subdivision of (L,ψ)).

<u>Proof</u>. By induction on depth($\underline{\underline{A}}$). If depth($\underline{\underline{A}}$) = 0, then $A = c(\underline{\underline{A}})$ and we can take $(\hat{K},\hat{\varphi}) = (K,\varphi)$. Assume now that depth($\underline{\underline{A}}$) > 0. By definition $(K^+,\varphi^+) = (K,\varphi)|c(\underline{\underline{A}}^+)$ is a regular triangulation of $c(\underline{\underline{A}}^+)$. By induction we can construct a canonical extension $(\hat{K}^+,\hat{\varphi}^+)$ of (K^+,φ^+); it is a smooth triangulation of $\underline{\underline{A}}^+$. By (2), $(\hat{K}^+,\hat{\varphi}^+)|A^0$ exists and is a canonical extension of $(K^0,\varphi^0) = (K^+,\varphi^+)|c(\underline{\underline{A}}^0)$; set $(\hat{K}^0,\hat{\varphi}^0) = (\hat{K}^+,\hat{\varphi}^+)|A^0$. Since the triangulation (K,φ) is regular, $c(p)$ is simplicial with respect to (K^0,φ^0) and $(K^*,\varphi^*) = (K,\varphi)|A^*$. By (3) and Lemma 9.6.2 there exist a subdivision $(\hat{K}^*,\varphi^*)$ of (K^*,φ^*) and a subdivi-

sion $(\tilde{K}^0,\hat{\varphi}^0)$ of $(\hat{K}^0,\hat{\varphi}^0)$ such that $p : A^0 \to A^*$ is simplicial with respect to $(\tilde{K}^0,\hat{\varphi}^0)$ and $(\hat{K}^*,\varphi^*)$. Let $s_p : \tilde{K}^0 \to \hat{K}^*$ be the corresponding simplicial map, i.e. $p = \varphi^* \circ |s_p| \circ (\hat{\varphi}^0)^{-1}$. Let K_{s_p} be the simplicial mapping cylinder of s_p and $\varphi_{s_p} : |K_{s_p}| \to C(p)$ be the homeomorphism constructed in 9.4.2. Clearly $(K_{s_p},\Phi\circ\varphi_{s_p})$ is a smooth triangulation of $\underline{\underline{A}}^-$. Choose now a subdivision $(\tilde{K}^+,\hat{\varphi}^+)$ of $(\hat{K}^+,\hat{\varphi}^+)$ which extends the subdivision $(K_{s_p},\Phi\circ\varphi_{s_p})|A^0$ of $(\hat{K}^0,\hat{\varphi}^0)$. Clearly $(\tilde{K}^+,\hat{\varphi}^+)$ and $(K_{s_p},\Phi\circ\varphi_{s_p})$ fit together and give rise to a triangulation $(\hat{K},\hat{\varphi})$ of A. It is easily seen that $(\hat{K},\hat{\varphi})$ is a smooth triangulation of $\underline{\underline{A}}$ and that it verifies (1) and (2). To check (3), use induction, the relation $f^0 = f^* \circ p$ and 9.6. Q.E.D.

7.8. THEOREM. Let $\underline{\underline{A}}$ be an a.s. of finite depth. Then there exists a smooth triangulation (K,φ) of $\underline{\underline{A}}$. If $f : \underline{\underline{A}} \longmapsto \underline{\underline{M}}$ is a proper submersive weak morphism, $\underline{\underline{M}}$ being of depth zero, and (L,ψ) is a good triangulation of M, then one can choose (K,φ) such that f is simplicial with respect to (K,φ) and a subdivision of (L,ψ).

Proof. By Theorem 6.5 there exists a total decomposition $\mathcal{D}$ of $\underline{\underline{A}}$. By Lemma 7.5 there exists a regular triangulation of $c(\underline{\underline{A}},\mathcal{D})$. The first assertion follows from Proposition 7.7. The second one also follows if the total decomposition $\mathcal{D}$ of $\underline{\underline{A}}$ is chosen f-compatible and the triangulation of $c(\underline{\underline{A}},\mathcal{D})$ is chosen such that $c(f)$ is simplicial with respect to it and a subdivision of (L,ψ). Q.E.D.

7.9. COROLLARY. Any subanalytic set is triangulable. In particular (real) analytic, semianalytic, algebraic or semialgebraic sets are triangulable.

Proof. By $[Hi_1]$ or $[Ha_1]$ any subanalytic set can be endowed with the structure of a Whitney stratification and then, by [Gib, Chap. II], with the structure of an a.s. Apply the above theorem. Q.E.D.

The notions and notation involved in the next corollary are standard, see for example [Br].

7.10. COROLLARY. Let M be a smooth manifold, let G be a compact Lie group and assume that G acts smoothly on M. Let M/G be the orbit space. Then, there exists a triangulation (K,φ) of M/G such that the image under φ of any open simplex of K consists of points of the same orbit type.

Proof. It is known that one can endow M/G with an a.s. structure such that the strata consist of points of the same orbit type. Apply 7.8.

Notes. By now, there are several published proofs of the triangulability of a.s.'s (or of stratified sets in Thom's sense) without boundary ([G_1], [J_2], [Ka], [Ve_3]) and also some unpublished ([He], [Lel] and [Mat]). As the referee informed me, Hendricks' approach is incomplete. Also, the proof given in [Lel] is incomplete: it is based on a result of Cairns which, as mentioned in the introduction of [Il], is not correct. The preprint of Matumoto is perhaps too sketchy: the essential difficulties (a structure result for stratified sets and a triangulability result for certain maps) were dismissed each in a sentence. Returning to the published proofs, Johnson settles the case of compact stratified sets in Thom's sense; its proof is short and elegant, but it is based on a very formalized structure theorem whose proof (given in [J_3]) is very long. The papers of Goresky and myself deal with a.s.'s and the approaches are more geometric. The proof given in this section combines ideas of these two papers.

A first attempt to triangulate orbit spaces is due to Yang [Y], but (see the introduction of [Il]) his proof is incomplete. Illman [Il] proves the triangulability of orbit spaces in the case of smooth actions of finite groups. Using the same arguments as in 7.10, the general case (smooth actions of compact Lie groups) was settled in [Ve_3]. The case of compact orbit spaces was also derived in [J_2] and [Mat].

First attempts to triangulate (real) algebraic and analytic sets are contained in [W], [Le], [K-B] and [L-W]. Rigorous proofs in the more general case of semianalytic sets are due to Lojasiewicz [Lo] and Giesecke [Gi]. The

The triangulability of subanalytic sets was first proved by Hironaka [Hi_1] and Hardt [Ha_2]. In the semialgebraic case simpler proofs are given in [Hi_1], [De] and [Cos]. All these triangulations are subanalytic (resp. semialgebraic).

8. TRIANGULATION OF NICE ABSTRACT THOM MAPPINGS

8.1. In this chapter $f : \underline{\underline{B}} \mapsto \underline{\underline{A}}$ will denote a proper a.T.m., $\underline{\underline{A}}$ and $\underline{\underline{B}}$ being of finite depth. We shall fix a regular total decomposition $\mathfrak{D} = \{\nabla^1, \ldots, \nabla^{n+1}\}$ of f, with $\mathcal{D}_{\mathfrak{D}} = \{\Delta^1, \ldots, \Delta^{n+1}\}$, $\nabla^1 = \{\Delta^1, \underline{\underline{B}}_-, \underline{\underline{B}}_+, S, \Psi\}$ and $\Delta^1 = \{\underline{\underline{A}}^-, \underline{\underline{A}}^+, p, \epsilon, \Phi\}$. $\mathfrak{D}_+$ (resp. $\mathfrak{D}_0, \mathfrak{D}_0^-, \mathfrak{D}_0^+, \ldots$) will always denote the regular total decomposition of $f_+ : \underline{\underline{B}}_+ \mapsto \underline{\underline{A}}^+$ (resp. $f_0 : \underline{\underline{B}}_0 \mapsto \underline{\underline{A}}^0$, $f_0^- : \underline{\underline{B}}_0^- \mapsto \underline{\underline{A}}^0$, $f_0^+ : \underline{\underline{B}}_0^+ \mapsto \underline{\underline{A}}^0, \ldots$) induced by $\mathfrak{D}$. On $\underline{\underline{A}}^+$ (resp. $\underline{\underline{A}}^0$, $\underline{\underline{B}}_*$, $\ldots$) we shall always consider the regular total decomposition $\mathcal{D}_{\mathfrak{D}_+}$ (resp. $\mathcal{D}_{\mathfrak{D}_0}$, $\mathcal{D}_S$, $\ldots$) induced by $\mathfrak{D}$. In order to simplify the notation we shall omit $\mathfrak{D}, \mathfrak{D}_+, \ldots, \mathcal{D}_{\mathfrak{D}}, \mathcal{D}_{\mathfrak{D}_+}, \ldots, \mathcal{D}_S, \ldots$ in expressions like $c(\underline{\underline{B}},\mathfrak{D})$, $c(f,\mathfrak{D})$, $c(\underline{\underline{B}}_+,\mathfrak{D}_+), \ldots, c(\underline{\underline{A}},\mathcal{D}_{\mathfrak{D}}), \ldots, c(\underline{\underline{B}}_*, \mathcal{D}_S), \ldots$ and denote them $c(\underline{\underline{B}})$, $c(f)$, $c(\underline{\underline{B}}_+), \ldots, c(\underline{\underline{A}}), \ldots, c(\underline{\underline{B}}_*), \ldots$ If B_i is a face of $\underline{\underline{B}}$, the same convention applies to the restriction f_i of f endowed with the regular total decomposition $\mathfrak{D}|B_i$.

8.2.1. Let $\{(L,\psi),(K,\varphi)\}$ be a good triangulation of $c(f)$. Since $c(\underline{\underline{B}})$ (resp. $c(\underline{\underline{A}})$) is the disjoint sum of $c(\underline{\underline{B}}_*)$ and $c(\underline{\underline{B}}_+)$ (resp. $\underline{\underline{A}}^*$ and $c(\underline{\underline{A}}^+)$), we can consider the good triangulations $\{(L_*, \psi_*),(K^*, \varphi^*)\} = \{(L,\psi),(K,\varphi)\}|c(f_*)$ and $\{(L_+,\psi_+),(K^+,\varphi^+)\} = \{(L,\psi),(K,\varphi)\}|c(f_+)$ of $c(f_*)$ and $c(f_+)$ respectively.

8.2.2. Let $\{(L,\psi),(K,\varphi)\}$ be a good triangulation of $c(f)$ and let B_i be a face of $\underline{\underline{B}}$. Then the restriction $\{(L,\psi),(K,\varphi)\}|c(f_i)$ exists and is a good triangulation of $c(f_i)$.

<u>Example</u>. With the notation introduced in **8.2.1**, we can consider $\{(L_+,\psi_+),(K^+,\varphi^+)\}|c(f_0)$. This good triangulation of $c(f_0)$ will be denoted $\{(L_0,\psi_0),(K^0,\varphi^0)\}$.

8.3.1. Consider an admissible square of depth m

$$(\Sigma)\qquad \begin{array}{ccc} \underline{\underline{B}} & \overset{\sigma}{\rightsquigarrow} & \underline{\underline{N}} \\ f \downarrow & & \downarrow g \\ \underline{\underline{A}} & \underset{\pi}{\longrightarrow} & \underline{\underline{M}} \end{array}$$

and assume that $\mathcal{D}$ is Σ-compatible (see 6.11.8; we shall use the notation introduced there). Let

$$(\Sigma_{\mathcal{D}})\qquad \begin{array}{ccc} c_\Sigma(\underline{\underline{B}}) = c_\Sigma(\underline{\underline{B}},\mathcal{D}) & \xrightarrow{c_\Sigma(\sigma)=c_\Sigma(\sigma,\mathcal{D})} & c(\underline{\underline{N}},\mathcal{D}_\Sigma) = c(\underline{\underline{N}}) \\ c_\Sigma(f) = c_\Sigma(f,\mathcal{D}) \downarrow & & \downarrow c(g) = c(g,\mathcal{D}_\Sigma) \\ c(\underline{\underline{A}}) & \xrightarrow[c(\pi)=c(\pi,\mathcal{D}_{\mathcal{D}})]{} & M \end{array}$$

be the square constructed in 6.14.

The good triangulations $\{(L,\psi),(K,\varphi)\}$ and $\{(F,\tau),(E,\rho)\}$ of $c(f)$ and $c(g)$ respectively are called Σ-compatible if there exists a good triangulation $\{(L_\Sigma,\psi_\Sigma),(K,\varphi)\}$ of $c_\Sigma(f)$ such that

(i) $\{(L_\Sigma,\psi_\Sigma),(F,\tau)\}$ is a triangulation of $c_\Sigma(\sigma)$ and $\{(K,\varphi),(E,\rho)\}$ is a triangulation of $c(\pi)$;

(ii) if $m = 0$, then $(L_\Sigma,\psi_\Sigma) = (L,\psi)$ (in this case $c_\Sigma(\underline{\underline{B}}) = c(\underline{\underline{B}})$!);

(iii) if $m > 0$, then $(L,\psi)|c(\underline{\underline{B}}^+,\mathcal{D}^+)$, $(L,\psi)|c(\underline{\underline{B}}^\times,\mathcal{D}^\times)$ and $(L,\psi)|c(\underline{\underline{B}}^-,\mathcal{D}^-)$ exist and

(iii$_1$) $\{(L,\psi)|c(\underline{\underline{B}}^+,\mathcal{D}^+),(K,\varphi)\}$ and $\{(F,\tau)|c(\underline{\underline{N}}^+,\mathcal{D}_{\Sigma^+}),(E,\rho)\}$ are Σ^+-compatible (this makes sense by induction);

(iii$_2$) $(L,\psi)|c(\underline{\underline{B}}^-,\mathcal{D}^-) = (L_\Sigma,\psi_\Sigma)|c(\underline{\underline{B}}^-,\mathcal{D}^-)$;

(iii$_3$) let $pr_1 : c(\underline{\underline{B}}^\times,\mathcal{D}^\times) \to c(\underline{\underline{B}}^h,\mathcal{D}^-|B^h)$ and $pr_2 : c(\underline{\underline{B}}^\times,\mathcal{D}^\times) \to [0,1]$ be given by $\theta_\Sigma^{-1}(b) = (pr_1(b),pr_2(b))$, $\theta_\Sigma : \underline{\underline{B}}^h \times [0,1] \dashrightarrow \underline{\underline{B}}^\times$ being the weak isomorphism associated to the square Σ (see 6.7.1(6));

by 6.14, $\theta_\Sigma(c(\underline{\underline{B}}^h, D^-|B^h) \times [0,1]) = c(\underline{\underline{B}}^\times, D^\times)$; then pr_1 (resp. pr_2) is simplicial with respect to $(L,\psi)|c(\underline{\underline{B}}^\times, D^\times)$ and $(L,\psi)|c(\underline{\underline{B}}^h, D^-|B^h)$ (resp. a linear triangulation of $[0,1]$).

<u>Remarks</u>. (1) $(L_\Sigma, \psi_\Sigma) = (L,\psi)|c_\Sigma(\underline{\underline{B}})$.

(2) Set $(L^\times, \psi^\times) = (L,\psi)|c(\underline{\underline{B}}^\times, D^\times)$ and $\tilde{\psi}^\times = (\theta_\Sigma^{-1}|c(\underline{\underline{B}}^\times, D^\times)) \circ \psi^\times$: $|L^\times| \to c(\underline{\underline{B}}^h, D^-|B^h) \times [0,1]$. Then $(L^\times, \tilde{\psi}^\times)$ is a good triangulation of $c(\underline{\underline{B}}^h, D^-|B^h) \times [0,1]$ and (iii_3) just says that $(L^\times, \tilde{\psi}^\times)$ is the product of $(L,\psi)|c(\underline{\underline{B}}^h, D^-|B^h)$ and of a linear triangulation (J,ν) of $[0,1]$, for some orders on L and J (see Lemma 9.2.4).

8.3.2. Let Σ be as above, let B_i be a face of $\underline{\underline{B}}$ and let $\{(L,\psi),(K,\varphi)\}$ and $\{(F,\tau),(E,\rho)\}$ be good triangulations of $c(f)$ and $c(g)$ respectively which are Σ-compatible. Then the good triangulations $\{(L,\psi),(K,\varphi)\}|c(f_i)$ and $\{(F,\tau),(E,\rho)\}|c(g_i)$ of $c(f_i)$ and $c(g_i)$ are $(\Sigma|B_i)$-compatible (see 6.7.3 for the definition of $\Sigma|B_i$).

8.3.3. A good triangulation $\{(L,\psi),(K,\varphi)\}$ of $c(f)$ is called <u>regular</u> if

(i) (K,φ) is a regular triangulation of $c(\underline{\underline{A}})$;

(ii) (L_*, ψ_*) is a regular triangulation of $c(\underline{\underline{B}}_*)$;

(iii) if $\operatorname{depth}(\underline{\underline{A}}) > 0$, the good triangulations $\{(L_0,\psi_0),(K^0,\varphi^0)\}$ of $c(f_0)$ and $\{(L_*,\psi_*),(K^*,\varphi^*)\}$ of $c(f_*)$ are S-compatible;

(iv) if $\operatorname{depth}(\underline{\underline{A}}) > 0$, the good triangulation $\{(L_+,\psi_+),(K^+,\varphi^+)\}$ of $c(f_+)$ is regular (this makes sense by induction).

8.3.4. Let $\{(L,\psi),(K,\varphi)\}$ be a regular triangulation of $c(f)$ and let B_i be a face of $\underline{\underline{B}}$. Then $\{(L,\psi),(K,\varphi)\}|c(f_i)$ is a regular triangulation of $c(f_i)$.

8.4. Let (K,φ) be a regular triangulation of $c(\underline{\underline{A}})$. It seems more or less plausible that there should exist a regular triangulation of $c(f)$ of the form $\{(L,\psi),(K',\varphi)\}$, K' being a subdivision of K. However, for certain technical reasons I cannot prove this fact for all a.T.m.'s, but only for the nice ones (see 8.4.1). Fortunately the class of nice a.T.m.'s is large enough to contain interesting examples (e.g. if B and A are semialgebraic open sets in Euclidean spaces, $f : B \to A$ is polynomial and the restriction of f to its critical set is both proper and finite to one, then one can endow B and A with w.a.s. structures $\underline{\underline{B}}$ and $\underline{\underline{A}}$ such that f becomes a nice a.T.m.)

8.4.1. The proper a.T.m. $f : \underline{\underline{B}} \mapsto \underline{\underline{A}}$ is called <u>nice</u> if

(i) $\mathrm{depth}(\underline{\underline{B}}_*) \leq 1$;

(ii) if $\mathrm{depth}(\underline{\underline{A}}) > 0$ and $\mathrm{depth}(\underline{\underline{B}}_*) = 1$, then $f_*^* : B_*^* \to A^*$ is finite to one and in the admissible square (of depth zero !)

(S^+)

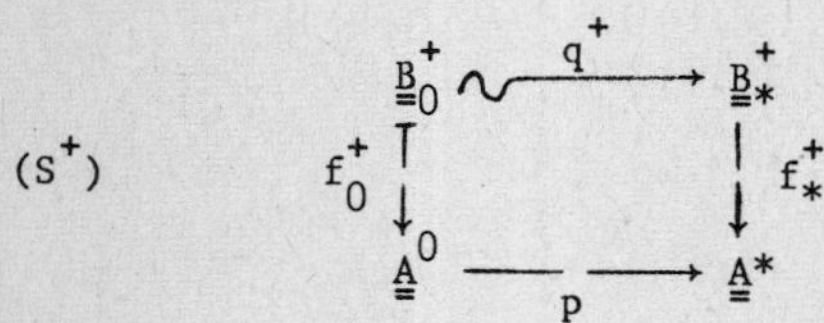

$\underline{\underline{B}}_0^+$ is the fibre product of $\underline{\underline{B}}_*^+$ and $\underline{\underline{A}}^0$ with respect to f_*^+ and p, and q^+ and f_0^+ are the canonical projections;

(iii) if $\mathrm{depth}(\underline{\underline{A}}) > 0$ and $\mathrm{depth}(\underline{\underline{B}}_*) = 0$, then either $f_* : B_* = B_*^* \to A^*$ is finite to one, or in the admissible square (of depth zero !)

(S)

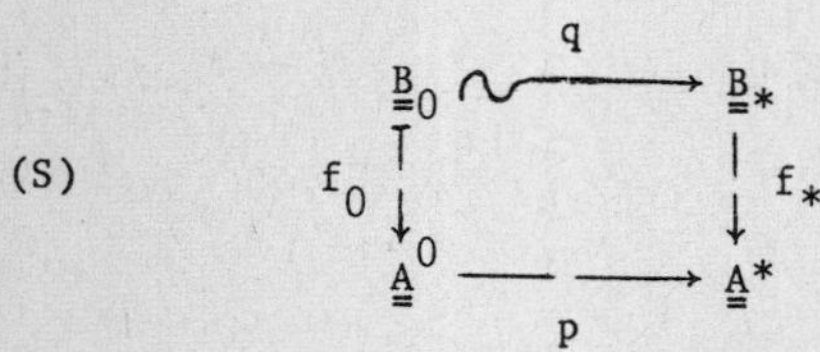

$\underline{\underline{B}}_0$ is the fibre product of $\underline{\underline{B}}_*$ and $\underline{\underline{A}}^0$ with respect to f_* and p, and q and f_0 are the canonical projections;

(iv) if $\mathrm{depth}(\underline{\underline{A}}) > 0$, then $f_+ : \underline{\underline{B}}_+ \mapsto \underline{\underline{A}}^+$ is nice (this makes sense by induction).

8.4.2. If f is nice and B_i is a face of $\underline{\underline{B}}$, then $f_i : \underline{\underline{B}}_i \mapsto \underline{\underline{A}}_i$ is also nice.

8.4.3. If $\underline{\underline{B}} = \underline{\underline{B}}' \cup_C \underline{\underline{B}}''$ and $f = f' \cup_C f''$, where $f' : \underline{\underline{B}}' \vdash\!\rightarrow \underline{\underline{A}}$ and $f'' : \underline{\underline{B}}'' \vdash\!\rightarrow \underline{\underline{A}}$ are a.T.m.'s, then f is nice if and only if f' and f'' are nice.

8.5. LEMMA. Let f be nice. Then $c(\underline{\underline{B}}_0^+) = c(\underline{\underline{A}}^0) \times_{A^*} B_*^+$. If $\{(L,\psi),(K,\varphi)\}$ is a regular triangulation of $c(f)$, then $(L,\psi)|c(\underline{\underline{B}}_0^+)$ is the fibre product of $(K^0,\varphi^0) = (K,\varphi)|c(\underline{\underline{A}}^0)$ and $(L,\psi)|B_*^+$ over $(K^*,\varphi^*) = (K,\varphi)|A^*$ (with respect to some orders).

Proof. The first assertion is an immediate consequence of 8.4.1 (iii) (notice that $\text{depth}(\underline{\underline{B}}_*^+) = 0$!). The second assertion follows from the first one, 8.3.3 (iii) and Lemma 9.3.4. Q.E.D.

8.6. LEMMA. Let f be nice, let $\{(L,\psi),(K,\varphi)\}$ be a regular triangulation of $c(f)$ and let (K',φ) be a subdivision of (K,φ). Let $I \subset I_{\underline{\underline{B}}}$ and for any $i \in I$ let $\{(L_i,\psi_i),(K_i,\varphi_i)\}$ be a regular triangulation of $c(f_i)$ such that (L_i,ψ_i) is a subdivision of (L,ψ), $(K_i,\varphi_i) = (K',\varphi)|c(\underline{\underline{A}}_i)$ and $(L_i,\psi_i)|c(B_i) \cap B_j = (L_j,\psi_j)|c(B_i) \cap B_j$ for any $i,j \in I$. Then there exists a subdivision (L',ψ) of (L,ψ) such that $\{(L',\psi),(K',\varphi)\}$ is a regular triangulation of $c(f)$ and $(L',\psi)|c(B_i) = (L_i,\psi_i)$ for any $i \in I$.

Since the proof of this lemma is more or less obvious, we omit it.

8.7. LEMMA. Let (K,φ) be a regular triangulation of $c(\underline{\underline{A}})$. Let $I \subset I_{\underline{\underline{B}}}$ and for any $i \in I$ let $\{(L_i,\psi_i),(K,\varphi)|c(A_i)\}$ be a regular triangulation of $c(f_i)$. Assume that $(L_i,\psi_i)|c(\underline{\underline{B}}_i) \cap B_j = (L_j,\psi_j)|c(\underline{\underline{B}}_i) \cap B_j$ for any $i,j \in I$. Then there exists a regular triangulation $\{(L,\psi),(\tilde{K},\varphi)\}$ of $c(f)$ such that

(1) $(\tilde{K},\varphi)$ is a subdivision of (K,φ) ;

(2) $(L,\psi)|c(\underline{\underline{B}}_i)$ is a subdivision of (L_i,ψ_i), $i \in I$.

Proof. If $\text{depth}(\underline{\underline{A}}) = 0$, the assertion follows from Lemma 7.5. We can therefore assume that $\text{depth}(\underline{\underline{A}}) > 0$ and that the assertion is true for any nice a.T.m. $f' : \underline{\underline{B}}' \vdash\!\rightarrow \underline{\underline{A}}'$ with $\text{depth}(\underline{\underline{B}}') < \infty$ and $\text{depth}(\underline{\underline{A}}') < \text{depth}(\underline{\underline{A}})$.

Step I. By Lemma 7.5 there exists a good triangulation $\{(L_*,\psi_*),(K^*,\varphi^*)\}$ of $c(f_*)$ such that (L_*,ψ_*) is a regular triangulation of $c(\underline{\underline{B}}_*)$, $(K^*,\varphi^*) \lhd (K,\varphi)|A^*$ and $(L_*,\psi_*)|c(\underline{\underline{B}}_*) \cap B_i \lhd (L_i,\psi_i)|c(\underline{\underline{B}}_*) \cap B_i$.

Step II. We can subdivide (if necessary) (K,φ) and (L_i,ψ_i) and assume that $(K^*,\varphi^*) = (K,\varphi)|A^*$ and $(L_*,\psi_*)|c(\underline{\underline{B}}_*) \cap B_i = (L_i,\psi_i)|c(\underline{\underline{B}}_*) \cap B_i$, $i \in I$.

Step III. Let $(K^0,\varphi^0) = (K,\varphi)|c(\underline{\underline{A}}^0)$ and $(L^+_*,\psi^+_*) = (L_*,\psi_*)|B^+_*$. Clearly $\{(K^0,\varphi^0),(K^*,\varphi^*)\}$ is a triangulation of $c(p) : c(\underline{\underline{A}}^0) \to A^*$ and $\{(L^+_*,\psi^+_*),(K^*,\varphi^*)\}$ is a triangulation of $f^+_* : B^+_* \to A^*$. Since $c(\underline{\underline{B}}^+_0) =$ $= c(\underline{\underline{A}}^0) \times_{A^*} B^+_*$ (cf. Lemma 8.5) we can consider the triangulation $(L^+_0,\psi^+_0) = (K^0,\varphi^0) \times_{A^*} (L^+_*,\psi^+_*)$ of $c(\underline{\underline{B}}^+_0)$ with respect to some orders on K^0, L^+_* and K^* (see Lemma 9.3.4). Moreover, these orders can be chosen such that $(L^+_0,\psi^+_0)|c(\underline{\underline{B}}^+_0) \cap B_i = (L_i,\psi_i)|c(\underline{\underline{B}}^+_0) \cap B_i$, $i \in I$. There is no difficulty in checking that $\{(L^+_0,\psi^+_0),(K^0,\varphi^0)\}$ is a regular triangulation of $c(f^+_0) : c(\underline{\underline{B}}^+_0) \to c(\underline{\underline{A}}^0)$.

Step IV. Set $(L^0_0,\psi^0_0) = (L^+_0,\psi^+_0)|c(\underline{\underline{B}}^0_0)$ and $L^h_0 = L^0_0$; define $\psi^h_0 :$ $|L^h_0| \to c(\underline{\underline{B}}^h_0)$ by $\psi^h_0(x) = \theta^{-1}_S(\psi^0_0(x))$. Then, obviously, $\{(L^h_0,\psi^h_0),(K^0,\varphi^0)\}$ is a regular triangulation of $c(f^h_0) : c(\underline{\underline{B}}^h_0) \to c(\underline{\underline{A}}^0)$ and, for any $i \in I$, $(L^h_0,\psi^h_0)|c(\underline{\underline{B}}^h_0) \cap B_i = (L_i,\psi_i)|c(\underline{\underline{B}}_i) \cap B^h_0$ (this last assertion follows from 8.3.1 (iii_3)). By induction there exists a regular triangulation $\{(L^-_0,\psi^-_0),(\hat{K}^0,\varphi^0)\}$ of $c(f^-_0) : c(\underline{\underline{B}}^-_0) \to c(\underline{\underline{A}}^0)$ such that $(\hat{K}^0,\varphi^0) \lhd (K^0,\varphi^0)$, $(L^-_0,\psi^-_0)|c(\underline{\underline{B}}^h_0) \lhd (L^h_0,\psi^h_0)$ and $(L^-_0,\psi^-_0)|c(\underline{\underline{B}}^h_0) \cap B_i \lhd (L_i,\psi_i)|c(\underline{\underline{B}}_i) \cap B^h_0$, $i \in I$. After subdividing (if necessary), we can assume that $(\hat{K}^0,\varphi^0) =$ $= (K^0,\varphi^0)$, $(L^-_0,\psi^-_0)|c(\underline{\underline{B}}^h_0) = (L^h_0,\psi^h_0)$ and $(L^-_0,\psi^-_0)|c(\underline{\underline{B}}^h_0) \cap B_i =$ $= (L_i,\psi_i)|c(\underline{\underline{B}}_i) \cap B^h_0$, $i \in I$.

Consider the commutative diagram

$$\begin{array}{ccc} |L^-_0| & \xrightarrow{\mu} & |L^*_*| \\ {\scriptstyle |s_{f^-_0}|}\downarrow & & \downarrow{\scriptstyle |s_{f^*_*}|} \\ |K^0| & \xrightarrow[|s_p|]{} & |K^*| \end{array}$$

where $(L^*_*,\psi^*_*) = (L_*,\psi_*)|B^*_*$ and $\mu = (\psi^*_*)^{-1} \circ q^- \circ \psi^-_0$. Since $|s_{f^*_*}|$ is finite to one (cf. 8.4.1), it is easily seen that $\mu = |s_{q^-}|$ for some simplicial map $s_{q^-} : L^- \to L$. It follows that $\{(L^-_0,\psi^-_0),(L^*_*,\psi^*_*)\}$ is a good triangulation of $c(q^-)$.

<u>Step V</u>. Let $(L^\times_0,\tilde\psi) = (L^h_0 \times J,\psi^h_0 \times \nu)$, where (J,ν) is a linear triangulation of $[0,1]$, the product being taken with respect to some orders on L^h_0 and J. Clearly $(L^\times_0,\tilde\psi)$ is a good triangulation of $c(\underline{\underline{B}}^h_0) \times [0,1]$ and therefore, if $\psi^\times_0 = (\theta_S|c(\underline{\underline{B}}^h_0) \times [0,1]) \circ \tilde\psi$, then $(L^\times_0,\psi^\times_0)$ is a good triangulation of $c(\underline{\underline{B}}^\times_0)$.

<u>Step VI</u>. Clearly (L^+_0,ψ^+_0), $(L^\times_0,\psi^\times_0)$ and (L^-_0,ψ^-_0) fit together and give rise to a good triangulation (L_0,ψ_0) of $c(\underline{\underline{B}}_0)$ with the following properties :

(VI_1) $\{(L_0,\psi_0),(K^0,\varphi^0)\}$ is a regular triangulation of $c(f_0)$;

(VI_2) $(L_0,\psi_0)|c(\underline{\underline{B}}_0) \cap B_i = (L_i,\psi_i)|c(\underline{\underline{B}}_0) \cap B_i$, $i \in I$.

<u>Step VII</u>. By induction there exists a regular triangulation $\{(L_+,\psi_+),(K^+,\varphi^+)\}$ of $c(f_+)$ such that $(L_+,\psi_+)|c(\underline{\underline{B}}_0) \triangleleft (L_0,\psi_0)$, $(K^+,\varphi^+) \triangleleft (K,\varphi)|c(\underline{\underline{A}}^+)$ and $(L_+,\psi_+)|c(\underline{\underline{B}}_+) \cap B_i \triangleleft (L_i,\psi_i)|c(\underline{\underline{B}}_i) \cap B_+$, $i \in I$.

<u>Step VIII</u>. By taking some well chosen subdivisions of (L_*,ψ_*), (L_+,ψ_+) and (K,φ) and fitting together the two first ones, we get the required triangulation $\{(L,\psi),(\tilde K,\varphi)\}$ of $c(f)$. Q.E.D.

8.8. PROPOSITION. Let $\{(L,\psi),(K,\varphi)\}$ be a regular triangulation of $c(f)$. Then there exists a canonical construction of a triangulation $\{(\hat L,\hat\psi),(\hat K,\hat\varphi)\}$ of f (such a triangulation of f is called a <u>canonical extension</u> of $\{(L,\psi),(K,\varphi)\}$) with the following properties:

(1) $(\hat L,\hat\psi)$ is a smooth triangulation of $\underline{\underline{B}}$ and $(\hat L,\hat\psi)|c(\underline{\underline{B}})$ is a subdivision of (L,ψ) ;

(2) $(\hat K,\hat\varphi)$ is a canonical extension of (K,φ) (in particular it is a smooth triangulation of $\underline{\underline{A}}$);

(3) if B_i is a face of $\underline{\underline{B}}$, then $\{(\hat L,\hat\psi),(\hat K,\hat\varphi)\}|f_i$ exists and is a canonical extension of $\{(L,\psi),(K,\varphi)\}|c(f_i)$;

(4) consider the submersive weak morphisms $\pi : \underline{\underline{A}} \dashrightarrow \underline{\underline{M}}$ and $g : \underline{\underline{N}} \dashrightarrow \underline{\underline{M}}$, $\underline{\underline{M}}$ and $\underline{\underline{N}}$ being of depth zero. Assume that $\underline{\underline{B}} = \underline{\underline{A}} \times_M \underline{\underline{N}}$ (with respect to π and g)), that $f : \underline{\underline{B}} \vdash\!\!\rightarrow \underline{\underline{A}}$ is the canonical projection and that $\mathcal{D}$ is Σ-compatible, Σ being the square

$$(\Sigma) \qquad \begin{array}{ccc} \underline{\underline{B}} & \overset{\sigma}{\rightsquigarrow} & \underline{\underline{N}} \\ f \downarrow & & \downarrow g \\ \underline{\underline{A}} & \underset{\pi}{\dashrightarrow} & \underline{\underline{M}} \end{array} \qquad \sigma(a,x) = x .$$

In view of 6.15, $c(\underline{\underline{B}}) = c(\underline{\underline{A}}) \times_M N$ (with respect to $c(\pi)$ and g) and $c(f)$ is the canonical projection. Assume further that (L,ψ) is the fibre product of (K,φ) and a good triangulation (F,τ) of N (over a good triangulation (E,ρ) of M). Then $(\hat{L},\hat{\psi})$ is the fibre product of $(\hat{K},\hat{\varphi})$ and (F,τ) over (E,ρ) (with respect to some orders).

(5) Assume that $f = f' \cup_C f'' : \underline{\underline{B}}' \cup_C \underline{\underline{B}}'' \vdash\!\!\rightarrow \underline{\underline{A}}$, $f' : \underline{\underline{B}}' \vdash\!\!\rightarrow \underline{\underline{A}}$ and $f'' : \underline{\underline{B}}'' \vdash\!\!\rightarrow \underline{\underline{A}}$ being nice a.T,m.'s and that $\mathcal{D} = \mathcal{D}' \cup_C \mathcal{D}''$, for some regular total decompositions $\mathcal{D}'$ and $\mathcal{D}''$ of f' and f'' respectively (see 6.11.4). Then, in view of 6.13, $c(\underline{\underline{B}}) = c(\underline{\underline{B}}') \cup_{c(\underline{\underline{C}})} c(\underline{\underline{B}}'')$. Assume further that $\{(L,\psi),(K,\varphi)\}|c(f')$ and $\{(L,\psi),(K,\varphi)\}|c(f'')$ exist and are regular triangulations of $c(f')$ and $c(f'')$ respectively. Then $\{(\hat{L},\hat{\psi}),(\hat{K},\hat{\varphi})\}|f'$ and $\{(\hat{L},\hat{\psi}),(\hat{K},\hat{\varphi})\}|f''$ exist and are canonical extensions of $\{(L,\psi),(K,\varphi)\}|c(f')$ and $\{(L,\psi),(K,\varphi)\}|c(f'')$ respectively.

<u>Proof</u>. By induction on $n = \operatorname{depth}(\underline{\underline{A}})$. In $n = 0$, the proposition follows from Proposition 7.7. Thus we may assume that $n > 0$ and that the proposition is true for any a.T.m. $f' : \underline{\underline{B}}' \vdash\!\!\rightarrow \underline{\underline{A}}'$ with $\operatorname{depth}(\underline{\underline{A}}') < n$. Assume also that $\operatorname{depth}(\underline{\underline{B}}_*) = 1$ (this is the most difficult case)

Let $\{(L_+,\psi_+),(K^+,\varphi^+)\} = \{(L,\psi),(K,\varphi)\}|c(f_+)$. By induction there exists a smooth triangulation $\{(\hat{L}_+,\hat{\psi}_+),(\hat{K}^+,\hat{\varphi}^+)\}$ of f_+ with the required properties. By (3), $\{(\hat{L}_0,\hat{\psi}_0),(\hat{K}^0,\hat{\varphi}^0)\} = \{(\hat{L}_+,\hat{\psi}_+),(\hat{K}^+,\hat{\varphi}^+)\}|f_0$ is a canonical extension of $\{(L,\psi),(K,\varphi)\}|c(f_0)$. By (5), $\{(\hat{L}_0^+,\hat{\psi}_0^+),(\hat{K}^0,\hat{\varphi}^0)\}$ =

$= \{(\hat{L}_0,\hat{\psi}_0),(\hat{K}^0,\hat{\varphi}^0)\}|f_0^+$, $\{(\hat{L}_0^{\times},\hat{\psi}_0^{\times}),(\hat{K}^0,\hat{\varphi}^0)\} = \{(\hat{L}_0,\hat{\psi}_0),(\hat{K}^0,\hat{\varphi}^0)\}|f_0^{\times}$ and $\{(\hat{L}_0^-,\hat{\psi}_0^-),(\hat{K}^0,\hat{\varphi}^0)\} = \{(\hat{L}_0,\hat{\psi}_0),(\hat{K}^0,\hat{\varphi}^0)\}|f_0^-$ exist and are canonical extensions of the corresponding restrictions of $\{(L,\psi),(K,\varphi)\}$. By (4), $(\hat{L}_0^+,\hat{\psi}_0^+)$ is the fibre product of $(\hat{K}^0,\hat{\varphi}^0)$ and $(L_*^+,\psi_*^+) = (L,\psi)|B_*^+$ over $(K^*,\varphi^*) = (K,\varphi)|A^*$, with respect to f_*^+ and p . Since $\Psi^+ : C(q^+) \to B_-^+$ is a homeomorphism, we can use 9.5.2 and obtain in an obvious way a smooth triangulation $\{(\tilde{L}_-^+,\tilde{\psi}_-^+),(\hat{K}^-,\hat{\varphi}^-)\}$ of $f_-^+ : B_-^+ \to A^-$ (to be more precise, $\hat{K}^-$ is the mapping cylinder of $s_p : \hat{K}^0 \to K^*$, $\hat{\varphi}^- = \Phi\circ\varphi_{s_p}$, $\tilde{L}_-^+$ is the fibre product of $\hat{K}^-$ and L_*^+ over K^* (with respect to the canonical retraction of $\hat{K}^-$ on K^* and $s_{f_*^+}$) and $\tilde{\psi}_-^+ = \Psi^+\circ\mu\circ\nu$, where $\nu : |\tilde{L}_-^+| \to C(p) \times_{A^*} B_*^+$ and $\mu : C(p) \times_{A^*} B_*^+ \longrightarrow C(q^+)$ are the homeomorphisms constructed in 9.3.4 and 9.5.1 respectively).

Using again (4), we see that $(\hat{L}_0^{\times},\hat{\psi}_0^{\times}) = (\hat{L}_0^{\times},\theta_S\circ\tilde{\psi}_0^{\times})$, where $(\hat{L}_0^{\times},\tilde{\psi}_0^{\times})$ is the product triangulation of $B_0^h \times [0,1]$ obtained from the triangulation $(\hat{L}_0^{\times},\hat{\psi}_0^{\times})|B_0^h$ of B_0^h and a linear triangulation of $[0,1]$. Since $\pi_{\nabla}|B_0^0 : B_0^0 \to B_*^*$ is simplicial with respect to $(\hat{L}_0^{\times},\hat{\psi}_0^{\times})|B_0^0$ and (L_*^*,ψ_*^*) , it follows that $\pi^{\times} = \pi_{\nabla}|B_0^{\times} : B_0^{\times} \longrightarrow B_*^*$ is simplicial with respect to $(\hat{L}_0^{\times},\hat{\psi}_0^{\times})$ and (L_*^*,ψ_*^*). Let $s_0^{\times} : \hat{L}_0^{\times} \longrightarrow L_*^*$ be the corresponding simplicial mapping.

From the construction of $(\tilde{L}_-^+,\tilde{\psi}_-^+)$ it also follows that $\pi_-^0 = \pi_{\nabla}|B_-^0 : B_-^0 \longrightarrow B_*^*$ is simplicial with respect to $(\tilde{L}_-^0,\tilde{\psi}_-^0) = (\tilde{L}_-^+,\tilde{\psi}_-^+)|B^0$ and (L_*^*,ψ_*^*) ; let $s_-^0 : \tilde{L}_-^0 \longrightarrow L_*^*$ be the corresponding simplicial mapping.

We can fit together $\tilde{L}_-^0$ and $\tilde{L}_0^{\times}$ along L_0^0 and obtain a new simplicial complex $L^{\vee}$; s_-^0 and $s_0^{\times}$ also fit together and give rise to a simplicial map $s^{\vee} : L^{\vee} \longrightarrow L_*^*$.

Set $B^{\vee} = B_-^0 \cup B_0^{\times}$ and let $\pi^{\vee} : B^{\vee} \to B_*^*$ be given by $\pi^{\vee}|B_-^0 = \pi_-^0$ and $\pi^{\vee}|B_0^{\times} = \pi_0^{\times}$. Define also $\psi^{\vee} : |L^{\vee}| \to B^{\vee}$ by $\psi^{\vee}\big||\tilde{L}_-^0| = \tilde{\psi}_-^0$ and $\psi^{\vee}\big||\tilde{L}_0^{\times}| = \tilde{\psi}_0^{\times}$. Clearly $(L^{\vee},\psi^{\vee})$ is a triangulation of $\pi^{\vee}$, the corresponding simplicial map being $s^{\vee}$.

Define a homeomorphism $\alpha : C(\pi^{\vee}) \longrightarrow B_-^{\times}$ as follows :

- if $[b,t] \in C(\pi^{\vee})$ with $b \in B_-^0$, then $b = \Psi([b_0,s])$ with $b_0 \in B_0^0$; let $b_h \in B_0^h$ be given by $\theta_S(b_h,1) = b_0$ and set

$$\alpha([b,t]) = \Psi([\theta_S(b_h,t),ts]) ;$$

- if $[b,t] \in C(\pi^\vee)$ with $b \in B_0^\times$, then $b = \theta_S(b_h,s)$ with $b_h \in B_0^h$; set

$$\alpha([b,t]) = \Psi([\theta_S(b_h,ts),t]) .$$

A direct verification shows that α is a homeomorphism. Define a triangulation $(L_-^\times,\psi_-^\times)$ of $B_-^\times$ by setting $L_-^\times = K_{s^\vee}$ (the mapping cylinder of $s^\vee$) and $\psi_-^\times = \alpha\circ\varphi_{s^\vee}$ ($\varphi_{s^\vee} : |K_{s^\vee}| \to C(\pi^\vee)$ is the homeomorphism constructed in 9.4.2).

Since $f_*^* : B_*^* \longrightarrow A^*$ is proper and finite to one, it is a covering and therefore $q^- : B_0^- \longrightarrow B_*^*$ is necessarily simplicial with respect to $(\hat{L}_0^-,\hat{\psi}_0^-)$ and (L_*^*,ψ_*^*) . Then $(\hat{L}_-^-,\hat{\psi}_-^-) = (K_{s_{q^-}},\Psi^-\circ\varphi_{s_{q^-}})$ is a triangulation of B_-^- and, if the barycentric subdivisions involved are chosen appropriately, $\{(\hat{L}_-^-,\hat{\psi}_-^-),(\hat{K},\hat{\varphi})|A^-\}$ is a triangulation of f_-^- . By subdividing $(\hat{L}_-^\times,\hat{\psi}_-^\times)$, $(\hat{L}_-^+,\hat{\psi}_-^+)$ and $(\hat{L}_+,\hat{\psi}_+)$, we can fit together these triangulations and $(\hat{L}_-^-,\hat{\psi}_-^-)$, and obtain a triangulation $(\hat{L},\hat{\psi})$ of B with the required properties. Q.E.D.

Combining the above results we obtain

8.9. THEOREM. Let $f : \underline{\underline{B}} \mapsto \underline{\underline{A}}$ be a proper and nice a.T.m., $\underline{\underline{B}}$ and $\underline{\underline{A}}$ being of finite depth. Then f is triangulable.

8.10. Consider now two smooth manifolds without boundary, M and N. We shall endow $C^\infty(M,N)$ with the (fine) C^∞-topology. A smooth map $f : M \to N$ is called <u>topologically stable</u> if there exists a neighborhood V of f in $C^\infty(M,N)$ such that for any $g \in V$ there exist homeomorphisms $\varphi : M \to M$ and $\psi : N \to N$ making commutative the diagram

$$\begin{array}{ccc} M & \xrightarrow{\varphi} & M \\ f\downarrow & & \downarrow g \\ N & \xrightarrow[\psi]{} & N \end{array} .$$

By definition, the subset $C^{\infty}_{t-st}(M,N)$ of topologically stable mappings is open in $C^{\infty}(M,N)$. Let $C_{pr}(M,N)$ denote the set of proper mappings from M to N. As conjectured by Thom and proved by Mather [Ma$_2$] (see also [Gib] the subset $C^{\infty}_{t-st}(M,N) \cap C_{pr}(M,N)$ is dense in $C^{\infty}(M,N) \cap C_{pr}(M,N)$. We shall prove now the following:

8.11. THEOREM. Let M and N be smooth manifolds without boundary. Then any proper, topologically stable smooth map $f : M \to N$ is triangulable.

<u>Proof</u>. Let $\Omega^k(M,N) \subset C^{\infty}_{t-st}(M,N) \cap C_{pr}(M,N)$ be the subset introduced in [Gib, Chapter IV, Proposition 4.1], k being sufficiently large.

<u>Step I</u>. We shall prove here that any $f \in \Omega^k(M,N)$ is triangulable. From (the proof of) Proposition 3.3 of Chapter IV in [Gib] it follows that there exists a Thom stratification (A,A') of f such that

(1) $A' = \{f(X);\ X \in A\} \cup \{N \setminus f(M)\}$;

(2) if $\Sigma \subset M$ denotes the set of critical points of f, then for any $Y \in A'$, $f^{-1}(Y) \cap \Sigma \in A$, $f^{-1}(Y) \setminus \Sigma \in A$ and the restriction of f to $f^{-1}(Y) \cap \Sigma$ is finite to one.

Let $T' = \{T_Y;\ Y \in A'\}$ be a controlled system of tubular neighborhoods of the Whitney stratification A' of N and let $T = \{T_X\ ;\ X \in A\}$ be a T'-controlled system of tubular neighborhoods of the Whitney stratification A of M. Let $\underline{\underline{M}}$ (resp. $\underline{\underline{N}}$) denote this w.a.s. (resp. a.s.) structure on M (resp. N) (see [Gib] for the notions used above).

Let $Y \in A'$ be such that $\emptyset \neq X = f^{-1}(Y) \setminus \Sigma \in A$. We can assume that $f(T_X) \subset T_Y$; then the diagram

$$\begin{array}{ccc} T_X & \xrightarrow{\pi_X} & X \\ {\scriptstyle f|T_X}\downarrow & & \downarrow{\scriptstyle f|X} \\ T_Y & \xrightarrow[\pi_Y]{} & Y \end{array}$$

is commutative and an easy dimension argument shows that the canonical map $\alpha : T_X \to T_Y \times_Y X$ (i.e. $\alpha(u) = (f(u),\pi_X(u))$) is a local diffeomorphism. Since $X = f^{-1}(Y) \cap (M \setminus \Sigma)$, it is obvious that $\alpha|X$ is a diffeomorphism onto $\alpha(X) = Y \times_Y X$. A standard argument (see for example [Go], pag. 150, Th. 3.3.1) shows that α is a diffeomorphism of a neighborhood of X on a neighborhood of $\alpha(X)$. But this implies immediately that f is a nice a.T.m. from $\underline{\underline{M}}$ to $\underline{\underline{N}}$. By Theorem 8.9 it is triangulable.

Step II. Let $f \in C^\infty_{t\text{-}st}(M,N) \cap C_{pr}(M,N)$ and let $V \subset C^\infty(M,N)$ be a neighborhood of f with the property required in the definition of a topologically stable map. From the proof of Corollary 4.5 in Chapter IV of [Gib] it follows that there exists $g \in \Omega^k(M,N) \cap V$. Let $\mu : M \to M$ and $\nu : N \to N$ be homeomorphisms such that $\nu \circ f = g \circ \mu$. By Step I there exists a triangulation $\{(K,\varphi),(L,\psi)\}$ of g. Then $\{(K,\mu^{-1}\circ\varphi),(L,\nu^{-1}\circ\psi)\}$ is a triangulation of f. Q.E.D.

Combining the above result with the fact that $C^\infty_{t\text{-}st}(M,N) \cap C_{pr}(M,N)$ is dense in $C^\infty(M,N) \cap C_{pr}(M,N)$ ([Ma_2], [Gib]), we obtain the following corollary which was (implicitely) conjectured in [T_1] :

8.12. COROLLARY. The set of proper smooth mappings from M to N which are triangulable is dense (even generic) in $C^\infty(M,N) \cap C_{pr}(M,N)$.

8.13. In [Ha_2] it is proved that any proper light subanalytic map is triangulable. Using other results of Hardt [Ha_1], one can show easily that such a map can be endowed with the structure of a proper nice a.T.m. From our Theorem 8.9 we reobtain the triangulability of proper light subanalytic maps.

8.14. It is perhaps worth mentionning the following remark due to Hironaka [Hi_2]. Let $f : X \to Y$ be a surjective complex-analytic map of connected complex-analytic manifolds. If f is triangulable, then it is flat.

8.15. Returning to Theorem 8.9, one would like to prove the following more general result: any proper a.T.m. $f : \underline{\underline{B}} \mapsto \underline{\underline{A}}$ with $\underline{\underline{B}}$ and $\underline{\underline{A}}$ of finite depth is triangulable. The main difficulty in proving this consists in the following. Let

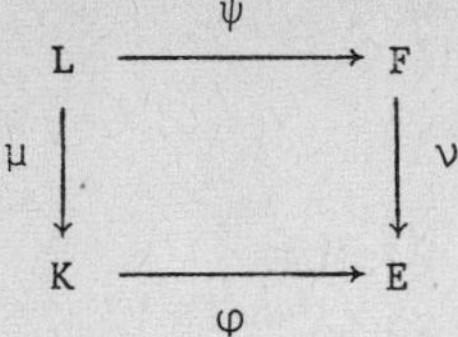

be a commutative diagram in the category of ordered simplicial complexes and increasing mappings. Assume that the canonical map $L \to K \times_E F$ is surjective. Let $p : C(|\psi|) \longrightarrow C(|\varphi|)$ denote the canonical map induced by $|\mu|$ and $|\nu|$. Does there exist a triangulation of p extending the given triangulations of $|\mu|$ and $|\nu|$? The answer is positive in two particular cases : (1) $|\nu|$ is finite to one, and (2) $L = K \times_E F$, ψ and μ being the canonical propjections. These cases appear in the proof of Theorem 8.9.

9. APPENDIX

9.1.1. We consider only countable and locally finite simplicial complexes (s.c.). Given a s.c. K we denote by $|K|$ its geometric realization and assume that $|K|$ is contained in some Euclidean space R^n , the inclusion $|K| \subset R^n$ being linear on the simplexes of K . If K' is a subdivision of K , we identify $|K'|$ and $|K|$ in the usual way.

By subcomplex we always mean full subcomplex and use the notation $L \triangleleft K$ to express the fact that L is a (full) subcomplex of K (thus if the vertices of a simplex σ of K are in L , then σ itself is in L). A simplex of K is also viewed, as usually, as a subcomplex of K . If $L \triangleleft K$, then $|L|$ is considered as a subspace of $|K|$, $|L| \subset |K|$, in the usual way.

A <u>triangulation</u> of a topological space A consists of a pair (K,φ), K being a s.c. and $\varphi : |K| \to A$ a homeomorphism.

If (K,φ) is a triangulation of A and K' is a subdivision of K, then, since $|K'| = |K|$, the pair (K',φ) is also a triangulation of A ; (K',φ) is called a <u>subdivision</u> of (K,φ) . If $A \subset R^n$ is a polyhedron, a triangulation (K,φ) of A is called <u>linear</u> if for any simplex σ of K

the restriction of φ to $|\sigma|$ is linear. For example $(K, 1_{|K|})$ is a linear triangulation of $|K|$.

Let (K,φ) be a triangulation of A and let $A_1 \subset A$ be a closed subset. We say that the restriction of (K,φ) to A_1 exists if $\varphi^{-1}(A_1) = |K_1|$ for some subcomplex K_1 of K ; if this is the case, then $(K_1, \varphi||K_1|)$ is a triangulation of A_1 , called the <u>restriction</u> of (K,φ) and denoted $(K,\varphi)|A_1$ (we shall also say that (K,φ) is an <u>extension</u> of $(K,\varphi||K_1|)$).

9.1.2. Given a simplicial map (s.m.) $s : K \to L$ we denote its geometric realization by $|s| : |K| \to |L|$. The s.m. is called <u>proper</u> if and only if the preimage of any vertex of L consists of finitely many vertices of K . Clearly s is proper if and only if $|s|$ is proper (i.e. the preimge through $|s|$ of any compact subset of L is compact).

A continuous map $f : B \to A$ is called <u>triangulable</u> if there exist triangulations (K,φ) and (L,ψ) of A and B respectively and a simplicial map $s_f : L \to K$ such that $f = \varphi \circ |s_f| \circ \psi^{-1}$ (we shall also say that f is <u>simplicial</u> with respect to (L,ψ) and (K,φ) , or that the pair $\{(L,\psi),(K,\varphi)\}$ is a <u>triangulation</u> of f); notice that s_f is completely determined by f , (K,φ) and (L,ψ) .

Let $\{(L,\psi),(K,\varphi)\}$ be a triangulation of $f : B \to A$ and let $B_1 \subset B$ and $A_1 \subset A$ be closed subsets such that $f(B_1) \subset A_1$. Set $f_1 = f|B_1 : B_1 \to A_1$. We say that the restriction of $\{(L,\psi),(K,\varphi)\}$ to f_1 exists if the restrictions of (L,ψ) to B_1 and (K,φ) to A_1 exist ; if this is so, then $\{(L,\psi)|B_1,(K,\varphi)|A_1\}$ is a triangulation of f_1 , called the <u>restriction</u> of $\{(L,\psi),(K,\varphi)\}$ and denoted $\{(L,\psi),(K,\varphi)\}|f_1$ (we shall also say that $\{(L,\psi),(K,\varphi)\}$ is an <u>extension</u> of $\{(L,\psi),(K,\varphi)\}|f_1$).

9.1.3. An <u>ordered simplicial complex</u> (o.s.c.) is a s.c. K such that any simplex σ of K is a totally ordered set, the order of any face τ of σ being induced by the order of σ .

Any s.c. can be endowed with a structure of an o.s.c. More generally, if $L \triangleleft K$ and L is endowed with a structure of an o.s.c., then this structure can be extended (the meaning is obvious) to a structure of an o.s.c. on K (the fact that L is a full subcomplex is essential here).

9.1.4. Let K be a s.c. and K' be a barycentric subdivision of K. (not necessarilly the standard one: the barycenter $\hat{\sigma}$ of a simplex σ of K may be any interior point of $|\sigma|$). On K' we shall always consider the following order: $\hat{\sigma} < \hat{\tau}$ if and only if σ is a face of τ (denote this by $\sigma \prec \tau$). Together with this order K' is an o.s.c.

9.2.1. Let K and L be o.s.c.'s. We shall construct a new o.s.c. $K \times L$, called the <u>product</u> of K and L, as follows: the set of vertices of $K \times L$ is the cartesian product of the set of vertices of K and the set of vertices of L; $((v_1,w_1),\ldots,(v_n,w_n))$ is a simplex of $K \times L$ if and only if $v_1,\ldots,v_n$ (resp. $w_1,\ldots,w_n$) are vertices of a simplex of K (resp. L) and $v_1 \leq v_2 \leq \ldots \leq v_n$ (resp. $w_1 \leq w_2 \leq \ldots \leq w_n$); if (v_1,w_1) and (v_2,w_2) are vertices of a simplex of $K \times L$ then $(v_1,w_1) \leq (v_2,w_2)$ if and only if $v_1 \leq v_2$ and $w_1 \leq w_2$.

Let $p_1 : K \times L \to K$ and $p_2 : K \times L \to L$ be the canonical projections (they are s.m.'s !). We can therefore consider $|p_1| \times |p_2| : |K \times L| \to |K| \times |L|$. The next lemma is well known (see for example [E-S], Chap. II, Lemma 8.9) .

9.2.2. LEMMA. $(K \times L, |p_1| \times |p_2|)$ is a triangulation of $|K| \times |L|$.

We shall also need the following result.

9.2.3. LEMMA. Let K and L be s.c.'s and let (N,ψ) be a triangulation of $|K| \times |L|$ such that $\{(N,\psi),(K,1_{|K|})\}$ (resp. $\{(N,\psi),(L,1_{|L|})\}$) is a triangulation of the projection $|K| \times |L| \longrightarrow |K|$ (resp. $|K| \times |L| \longrightarrow |L|$). Then K and L can be endowed with structures of o.s.c.'s such that

(1) there exists an isomorphism $\mu : K \times L \to N$;

(2) $\psi \circ |\mu| = |p_1| \times |p_2|$, p_1 and p_2 being as in 9.2.2.

(3) if $\dim(K) \geq 1$ and $\dim(L) \geq 1$, then the orders on K and L with the above properties are unique up to a simultaneous change to the oposite orders.

<u>Proof</u>. Assume that $\dim(K) \geq 1$ and $\dim(L) \geq 1$, the other cases being trivial. Let $\sigma = (u_1,u_2)$ be a 1-dimensional simplex of K. Set $u_1 \leq u_2$. Let $\tau = (v_1,v_2)$ be a 1-dimensional simplex of L. From the hypotheses it follows easily that $\psi^{-1}(|\sigma| \times |\tau|)$ is one of the following polyhedra

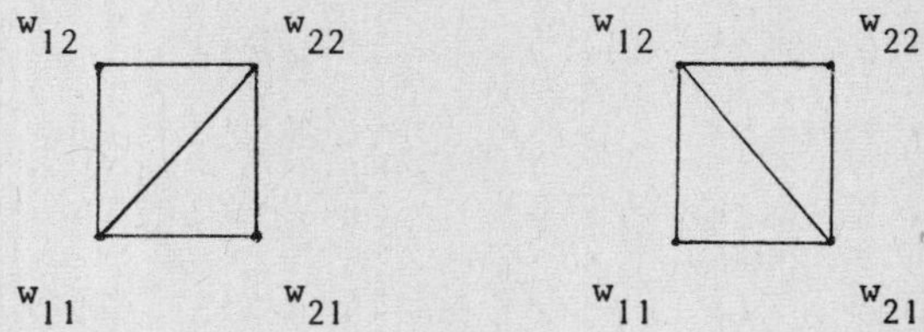

and $\psi(w_{ij}) = (u_i,v_j)$. In the first case set $v_1 < v_2$, while in the second case set $v_2 < v_1$. One can verify that this procedure determines a structure of o.s.c. on L. Next, starting with $\tau = (v_1,v_2)$ and with the order already determined on $\{v_1,v_2\}$, we endow K with a structure of o.s.c. The lemma follows without difficulty. Q.E.D.

The next lemma is now obvious.

9.2.4. LEMMA. Let (K,φ) and (L,ψ) be triangulations of A and B respectively. If K and L are endowed with structures of o.s.c.'s, then $(K \times L,\theta)$, with $\theta : |K \times L| \to A \times B$ given by $\theta(z) = (\varphi(|p_1|(z)), \psi(|p_2|(z)))$, is a triangulation of $A \times B$, called the <u>product</u> of (K,φ) and (L,ψ), and denoted $(K,\varphi) \times (L,\psi)$; the projections $A \times B \longrightarrow A$ and $A \times B \longrightarrow B$ are simplicial with respect to this triangulation of $A \times B$ and the given triangulations of A and B. Any triangulation of $A \times B$ with the last property is (up to an isomorphism) of the form $(K,\varphi) \times (L,\psi)$, for some orders on K and L.

9.3.1. Let K , L and M be o.s.c.'s and let $s : K \to M$ and $t : L \to M$ be increasing simplicial mappings. Let $K \times_M L$ be the full subcomplex of $K \times L$ whose vertices are all the vertices of $K \times L$ of the form (u,v) with $s(u) = t(v)$. $K \times_M L$ is called the <u>fibre product</u> of K and L over M (with respect to s and t). Let $\bar{p}_1 : K \times_M L \longrightarrow K$ and $\bar{p}_2 : K \times_M L \longrightarrow L$ denote the restrictions of the projections $p_1 : K \times L \longrightarrow K$ and $p_2 : K \times L \longrightarrow L$ respectively; define $p : K \times_M L \longrightarrow M$ by setting $p = s \circ \bar{p}_1 = t \circ \bar{p}_2$.

Consider now the (topological) fibre product $|K| \times_{|M|} |L| = \{(x,y) \in |K| \times |L| \ ; \ |s|(x) = |t|(y)\}$. One checks easily that $|\bar{p}_1| \times_{|M|} |\bar{p}_2| : |K \times_M L| \longrightarrow |K| \times_{|M|} |L|$ given by $(|\bar{p}_1| \times_{|M|} |\bar{p}_2|)((z) = (|\bar{p}_1|(z), |\bar{p}_2|(z)) = (|p_1|(z), |p_2|(z))$ is well defined and that $(|p_1| \times |p_2|)^{-1}(|K| \times_{|M|} |L|) = |K \times_M L|$. From Lemma 9.2.2 we get

9.3.2. LEMMA. $(K \times_M L, |\bar{p}_1| \times_{|M|} |\bar{p}_2|)$ is a triangulation of $|K| \times_{|M|} |L|$.

We shall also need the following generalization of Lemma 9.2.3.

9.3.3. LEMMA. Let K , L and M be s.c.'s, let $s : K \to M$ and $t : L \to M$ be s.m.'s and let (N,ψ) be a triangulation of $|K| \times_{|M|} |L|$ such that $\{(N,\psi),(K,1_{|K|})\}$ (resp. $\{(N,\psi),(L,1_{|L|})\}$) is a triangulation of the canonical projection $|K| \times_{|M|} |L| \longrightarrow |K|$ (resp. $|K| \times_{|M|} |L| \longrightarrow |L|$). Then we can endow K , L and M with structures of o.s.c.'s such that

(1) s and t are increasing with respect to these orders;

(2) there exists an isomorphism $\mu : K \times_M L \to N$;

(3) $\psi \circ |\mu| = |\bar{p}_1| \times_{|M|} |\bar{p}_2|$.

<u>Proof</u>. Choose any structure of o.s.c. on M . Given a vertex w of M let K_w be the (full) subcomplex of K whose vertices are all the vertices u of K with $s(u) = w$. Define similarly $L_w \triangleleft L$ (with respect to t). Notice that $\{(N,\psi),(M,1_{|M|})\}$ is a triangulation of the map $|p| \circ \psi : |N| \to |M|$; let $r : N \to M$ be the corresponding simplicial map. As above

we can define $N_w \triangleleft N$ (with respect to r). From the hypotheses it follows that $\psi(|N_w|) = |K_w| \times |L_w| \subset |K| \times_{|M|} |L|$ and therefore $(N_w, \psi| |N_w|)$ is a triangulation of $|K_w| \times |L_w|$. We can apply Lemma 9.2.3 and endow K_w and L_w with structures of o.s.c.'s with the properties stated there. Together with assertion (1) in the present lemma, these structures determine the required structures of o.s.c.'s on K and L. The remaining verifications are left to the reader. Q.E.D.

The next lemma is now obvious.

9.3.4. LEMMA. Let (K,φ), (L,ψ) and (M,ξ) be triangulations of A, B and C. Let $f : A \to C$ and $g : B \to C$ be simplicial with respect to these triangulations and let $s_f : K \to M$ and $s_g : L \to M$ be the corresponding s.m.'s. If K, L and M are endowed with structures of o.s.c.'s such that s_f ans s_g are increasing, then $(K \times_M L, \theta)$, with $\theta : |K \times_M L| \longrightarrow A \times_C B$ given by $\theta(z) = (\varphi(|\bar{p}_1|(z)), \psi(|\bar{p}_2|(z)))$, is a triangulation of $A \times_C B$, called the <u>fibre product</u> of (K,φ) and (L,ψ) over C (with respect to f and g) and denoted $(K,\varphi) \times_C (L,\psi)$; the projections $A \times_C B \longrightarrow A$ and $A \times_C B \longrightarrow B$ are simplicial with respect to this triangulation of $A \times_C B$ and the given triangulations of A and B. Any triangulation of $A \times_C B$ with the last property is, up to an isomorphism, of the form $(K,\varphi) \times_C (L,\psi)$, for some orders on K, L and M.

9.4.1. Let $s : K \to L$ be a proper s.m. Given a barycentric subdivision L' of L we can always find a barycentric subdivision K' of K such that s extends to a s.m. $s' : K' \to L'$ and $|s'| = |s|$. For any simplex σ we denote by $\hat{\sigma}$ the corresponding vertex of the barycentric subdivision.

We recall first the definition of the <u>simplicial mapping cylinder</u> M_s of s. The set of vertices of M_s is the disjoint union of the sets of vertices of L and K'. A finite set $(v_1,\dots,v_m,\hat{\sigma}_1,\dots,\hat{\sigma}_n)$ is a simplex of M_s if $\tau = (v_1,\dots,v_m)$ is a simplex of L and $\hat{\tau} \leq s'(\hat{\sigma}_1)$, $\hat{\sigma}_1 < \hat{\sigma}_2 < \dots < \hat{\sigma}_n$ (recall that $\hat{\sigma}' < \hat{\sigma}''$ means that σ' is a face of

σ''). In the above definition the v_i's (or the $\hat{\sigma}_i$'s) may be omitted, with the obvious changes in the defining conditions. Thus a simplex of M_s is of the form $\tau * A$ (the join of τ and A), τ being a simplex of L , A being a simplex of K' , $A = (\hat{\sigma}_1,\ldots,\hat{\sigma}_n)$ and $\hat{\tau} < s'(\hat{\sigma}_1)$. It is therefore clear that M_s is a subcomplex of the s.c. $L * K'$, the join of L and K' . There exists a canonical subdivision M_s' of M_s defined as follows: the set of vertices of M_s' is the disjoint union of the sets of vertices of L' and K' , and a set $(\hat{\tau}_1,\ldots,\hat{\tau}_m,\hat{\sigma}_1,\ldots,\hat{\sigma}_n)$ of vertices of M_s' is a simplex of M_s' if $\hat{\tau}_1 < \ldots < \hat{\tau}_m \leq s'(\hat{\sigma}_1)$ and $\hat{\sigma}_1 < \ldots < \hat{\sigma}_n$. Thus M_s' is a subcomplex of $L' * K'$.

There exist canonical s.m.'s $r_s : M_s' \to L'$, $i_s : L \to M_s$, $i_s' : L' \to M_s'$, $j_s : K' \to M_s$ and $j_s' : K' \to M_s'$ defined by $r_s(\hat{\tau}) = \hat{\tau} = i_s'(\hat{\tau})$ if τ is a simplex of L, $r_s(\hat{\sigma}) = s(\hat{\sigma}) = s'(\hat{\sigma})$, $j_s(\hat{\sigma}) = j_s'(\hat{\sigma}) = \hat{\sigma}$ if σ is a simplex of K and $i_s(v) = v$ if v is a vertex of L . We shall use i_s (resp. j_s) to identify L (resp. K') with a simplicial subcomplex of M_s .

Consider now the <u>topological mapping cylinder</u> $C_{|s|}$ of $|s| : |K| \to |L|$. Recall that $C_{|s|}$ is obtained from $(|K| \times [0,1]) \sqcup (|L| \times \{0\})$ (disjoint union) by identifying $(x,0) \in |K| \times [0,1]$ with $(|s|(x),0) \in |L| \times \{0\}$. The image of $(z,t) \in (|K| \times [0,1]) \sqcup (|L| \times \{0\})$ in $C_{|s|}$ is denoted $[z,t]$. Define $r_{|s|} : C_{|s|} \to |L|$, $i_{|s|} : |L| \to C_{|s|}$ and $j_{|s|} : |K| \to C_{|s|}$ by setting $r_{|s|}([x,t]) = |s|(x)$, $j_{|s|}(x) = [x,1]$ if $x \in |K|$ and $t \in [0,1]$ and $r_{|s|}([y,0]) = y$, $i_{|s|}(y) = [y,0]$ if $y \in |L|$. We shall use $i_{|s|}$ (resp. $j_{|s|}$) to identify $|L|$ (resp. $|K|$) with a closed subspace of $C_{|s|}$.

We shall prove now that there exists a triangulation of $C_{|s|}$ of the form (M_s', φ) . We shall follow Goresky's approach (see Section 4 of [G_1]) since the map φ he constructs has a smoothness property (see 9.4.2 (4) below) which is necessary in obtaining smooth triangulations of a.s.'s.

9.4.2. PROPOSITION. Let $s\colon K \to L$ be a proper s.m. Then there exists a homeomorphism $\varphi_s : |M_s| \to C_{|s|}$ with the following properties:

(1) If $K_1 \lhd K$ and $L_1 \lhd L$ are subcomplexes, $s(K_1) \lhd L_1$ and $s_1 : K_1 \to L_1$ is the restriction of s, then $\varphi_s(|M_{s_1}|) = C_{|s_1|}$ and $\varphi_s| \, |M_{s_1}| = \varphi_{s_1}$ (of course, the barycentric subdivisions of K_1 and L_1 are those induced by K' and L').

(2) $j_{|s|} = \varphi_s \circ |j_s|$, $i_{|s|} = \varphi_s \circ |i_s|$, $|r_s| = r_{|s|} \circ \varphi_s$.

(3) Let J be a s.c. and let $g\colon K' \to J$ and $h\colon L' \to J$ be s.m.'s such that $g = h \circ s'$. Define $\psi\colon C_{|s|} \to |J|$ and $s_\psi : M'_s \to J$ by setting $\psi([x,t]) = |g|(x)$ for $x \in |K'| = |K|$, $\psi([y,0]) = |h|(y)$ for $y \in |L'| = |L|$, $s_\psi(\hat{\tau}) = h(\hat{\tau})$ for $\hat{\tau} \in L'$ and $s_\psi(\hat{\sigma}) = g(\hat{\sigma})$ for $\hat{\sigma} \in K'$. Then $\{(M'_s, \varphi_s), (J, 1_{|J|})\}$ is a triangulation of ψ, the corresponding s.m. being s_ψ.

(4) The restiction $\varphi_s| \, |M_s| \smallsetminus |L| : |M_s| \smallsetminus |L| \to C_{|s|} \smallsetminus |L| = |K| \times (0,1]$ of φ_s is smooth (see [Mu], Section 8.1).

<u>Proof</u>. Let $\sigma = (u_1,\dots,u_m)$ be a simplex of K, $\tau = (v_1,\dots,v_n)$ be a simplex of L and assume that $s(\sigma) \preccurlyeq \tau$. Given $x = \sum \alpha_i u_i \in |\sigma|$ and $1 \le j \le n$ set $\alpha^j = \sum_{s(u_i)=v_j} \alpha_i$ and, if $\alpha^j > 0$, $x^j = \sum_{s(u_i)=v_j} (\alpha_i/\alpha_j)u_i$.

Let now $\theta * A$ be a simplex of M_s with $A = (\hat{\sigma}_1, \dots, \hat{\sigma}_r)$, $\theta \preccurlyeq s(\sigma_1)$ and $\hat{\sigma}_1 < \hat{\sigma}_2 < \dots < \hat{\sigma}_r$ (thus $\theta \preccurlyeq \tau$); let $z \in |\theta * A| \subset |M_s|$. Then there exist $t \in [0,1]$, $x \in |A| \subset |\sigma|$ and $y = \sum \beta_j v_j \in |\theta|$ such that $z = tx + (1-t)y$ (if $t \neq 0$ and $t \neq 1$, then x, y and t are unique). Since $\theta < s(\sigma_1)$, it follows that for any vertex $v_j \in \theta$, $\alpha^j \neq 0$ and we can define $x_y \in |\sigma|$ by setting $x_y = \sum_{v_j \in \theta} \beta_j x^j$. Define next $\varphi_{s,\sigma,\tau}(z) \in C_{|s|}$ by

$$\varphi_{s,\sigma,\tau}(z) = [tx + (1-t)x_y, t]$$

(notice that $|s|(x_y) = y$; thus when $t = 0$, $\varphi_{s,\sigma,\tau}(z) = [x_y,0] = [y,0]$ is still well defined; when $t = 1$, $\varphi_{s,\sigma,\tau}(z) = [x,1]$ is also well defined).

If $\tilde{\sigma}$, $\tilde{\tau}$, $\tilde{\theta}$ and $\tilde{A}$ are other simplexes satisfying the same conditions as σ, τ, θ and A and if $z \in |\tilde{\theta} * \tilde{A}|$, then one can check that $\varphi_{s,\sigma,\tau}(z) = \varphi_{s,\tilde{\sigma},\tilde{\tau}}(z)$. We can therefore define $\varphi_s : |M_s| \to C_{|s|}$ by $\varphi_s(z) = \varphi_{s,\sigma,\tau}(z)$, z , σ , τ and A being as above.

Assertions (1), (2), (3) and the continuity of φ_s are easy to verify. In order to prove that φ_s is a homeomorphism it is sufficient (by (1)) to consider the case $K = \sigma$ and $L = \tau = s(\sigma)$. If $\sigma = \tau$ and $s = 1_\sigma$, denote $M_s = M_\sigma$, $C_{|s|} = C_{|\sigma|}$ and $\varphi_s = \varphi_\sigma$. In this case $C_{|\sigma|} = |\sigma| \times [0,1] = |\sigma| \times |\Delta^1|$ (Δ^1 is the standard 1-dimensional simplex with vertices 0 and 1), $M_\sigma = \sigma' \times \Delta^1$ (σ' is considered with the order defined in 9.1.4 and Δ^1 with the order $0 < 1$) and $\varphi_\sigma : |\sigma' \times \Delta^1| \to |\sigma'| \times |\Delta^1|$ is the map determined by the projections of $\sigma' \times \Delta^1$ on σ' and Δ^1 . By Lemma 9.2.2 it is a homeomorphism.

Return now to the general case $s : \sigma \to \tau = s(\sigma)$. Let $\sigma = (u_1,\ldots, u_m)$ and $\tau = (v_1,\ldots,v_n)$. For $x \in |s|^{-1}(\hat{\tau})$ define $s_x : |\tau| \to |\sigma|$ by

$$s_x(y) = \Sigma\, y_j x^j \ , \quad y = \Sigma\, y_j v_j \in |\tau| \ .$$

It is clear that

$$(9.4.2.1) \qquad |s| \circ s_x = 1_{|\sigma|} \ .$$

Define also $F_x : |M_\tau| \to |M_s|$ and $G_x : C_{|\tau|} \to C_{|s|}$ as follows. Let $z \in |\theta * (\hat{\tau}_1,\ldots,\hat{\tau}_r)| \subset |M_\tau| = |M_{1_\tau}|$ (where $\theta < 1_\tau(\tau_1) = \tau_1 < \ldots < \tau_r < \tau$), $z = ty + \Sigma t_k \hat{\tau}_k$ with $y \in |\theta|$ and $t + \Sigma\, t_k = 1$. Let $\sigma_i = s^{-1}(\tau_i)$ and let A be the simplex of σ' generated by $\hat{\sigma}_1 < \ldots < \hat{\sigma}_r$. Then $\theta * A$ is a simplex of M_s and we can define $F_x(z) \in |\theta * A| \subset |M_s|$ by setting

$$F_x(z) = ty + \Sigma\, t_k s_x(\hat{\tau}_k) \ .$$

In view of (9.4.2.1) the definition is correct. G_x is defined by

$$G_x([y,t]) = [s_x(y),t] \ , \quad [y,t] \in C_{|\tau|} \ .$$

There are also canonical projections $\pi_s : M_s \to M_\tau$ and $p_s : C_{|s|} \to C_{|\tau|}$ given by

$$\pi_s(v_j) = v_j \quad \text{and} \quad \pi_s(\hat{\sigma}_0) = s(\hat{\sigma}_0) \quad \text{for} \quad \hat{\sigma}_0 < \sigma$$
$$p_s([x,t]) = [|s|(x),t] \,, \quad [x,t] \in C_{|s|} \,.$$

A direct verification shows that the diagram

(9.4.2.2)
$$\begin{array}{ccccc} |M_\tau| & \xrightarrow{F_x} & |M_s| & \xrightarrow{|\pi_s|} & |M_\tau| \\ \varphi_\tau \downarrow & & \downarrow \varphi_s & & \downarrow \varphi_\tau \\ C_{|\tau|} & \xrightarrow[G_x]{} & C_{|s|} & \xrightarrow[|p_s|]{} & C_{|\tau|} \end{array}$$

is commutative. From (9.4.2.1) it follows that

(9.4.2.3) $$|\pi_s| \circ F_x = 1_{|M_\tau|}$$

and

(9.4.2.4) $$p_s \circ G_x = 1_{C_{|\tau|}} \,.$$

Since $|\sigma| = \bigcup s_x(|\tau|)$, x running in $|s|^{-1}(\hat{\tau})$, and since $s_x(|\tau|) \cap s_{x'}(|\tau|) \cap |\sigma|^\circ = \emptyset$ if $x \neq x'$ ($|\sigma|^\circ$ is the interior of σ), the commutativity of the diagram (9.4.2.2), the relations (9.4.2.3) and (9.4.2.4) and the fact that φ_τ is a homeomorphism prove that φ_s is a homeomorphism too. Q.E.D.

9.4.3. LEMMA. Consider the following commutative diagram of s.m.'s

$$\begin{array}{ccc} K & \xrightarrow{s} & L \\ \alpha \downarrow & & \downarrow \beta \\ M & \xrightarrow[t]{} & N \end{array}$$

and assume that s and t are proper and $|\beta| : |L| \to |N|$ is finite to one. Let N' be a barycentric subdivision of N and chose barycentric subdivisions M' , K' and L' of M , K and L respectively such that

t, β, α and s extend to s.m.'s $t' : M' \to N'$, $\beta' : L' \to N'$, $\alpha' : K' \to M'$ and $s' : K' \to L'$ and $|t| = |t'|$, $|\beta| = |\beta'|$, $|\alpha| = |\alpha'|$ and $|s| = |s'|$ (this is always possible: one choses first M' and then K' such that $|t'| = |t|$ and $|\alpha| = |\alpha'|$; L' is uniquely determined by the condition $|\beta'| = |\beta|$). Define $s_\psi : M_s \to M_t$ and $\psi : C_{|s|} \to C_{|t|}$ by setting

$$s_\psi(v) = \beta(v) \text{ for any vertex } v \text{ of } L ;$$

$$s_\psi(\hat{\sigma}) = \alpha'(\hat{\sigma}) \text{ for any simplex } \sigma \text{ of } K ;$$

$$\psi([x,t]) = [|\alpha|(x),t] \text{ for } [x,t] \in C_{|s|} \text{ with } x \in |K| ;$$

$$\psi([y,0]) = [|\beta|(y),0] \text{ for } [y,0] \in C_{|s|} \text{ with } y \in |L| .$$

Then $\{(M_s,\varphi_s),(M_t,\varphi_t)\}$ is a triangulation of ψ, the corresponding s.m. being s_ψ.

<u>Proof</u>. Using (1) of Lemma 9.4.2 and the fact that $|\beta|$ is finite to one, it is sufficient to consider the case $K = \sigma$, $L = N = \tau$, $M = \theta$ and $\beta = 1_\tau$. This case can be settled by a direct verification. Q.E.D.

9.5.1. LEMMA. Let $f : A \to B$ be a proper continuous map and let $\beta : B_1 \to B$ be continuous. Let $A_1 = A \times_B B_1$ (with respect to f and β) and let $f_1 : A_1 \to B_1$ and $\alpha : A_1 \to A$ be the canonical projections. Let $\gamma : C_{f_1} \to C_f$ be the continuous map induced by α and β and let $r_f : C_f \to B$ and $r_{f_1} : C_{f_1} \to B_1$ be the canonical retractions. Define $h : C_{f_1} \to C_f \times_B B_1$ by $h(z) = (\gamma(z),r_{f_1}(z))$ (the fibre product is taken with respect to r_f and β). Then h is a homeomorphism and $h^{-1}\circ\gamma$ and $h^{-1}\circ r_{f_1}$ are the canonical projections of $C_f \times_B B_1$ on C_f and B_1 respectively.

The <u>proof</u> of this lemma is simple and left to the reader.

9.5.2. LEMMA. The notation is as above. Assume that f and β are triangulable with respect to some triangulations of A and B_1 and the same triangulation of B. Then γ is triangulable.

Proof. There is no loss of generality in assuming that $A = |K|$, $B_1 = |L_1|$, $B = |L|$, $f = |s|$ and $\beta = |t|$. By subdividing first L and then K and L_1, we may further assume that $A = |K'|$, $B = |L'|$, $B_1 = |L_1'|$, $f = |s'|$ and $\beta = |t'|$. We can therefore consider $K_1 = K' \times_{L'} L_1'$ and, by 9.3.2, we can identify A_1 with $|K_1|$; under this identification $\alpha = |\bar{p}_1|$ and $f_1 = |\bar{p}_2|$, $\bar{p}_1$ and $\bar{p}_2$ being the projections of K_1 on K' and L_1'. Let $r_s : M_s' \to L'$ be the retraction. Since M_s' has an obvious structure of o.s.c., we can consider $M = M_s' \times_{L'} L_1'$, with respect to r_s and t'. Using the homeomorphisms $|\tilde{M}| \to |M_s'| \times_{|L|} |L_1'|$ of 9.3.2 and $|M_s'| \to C_{|s|} = C_f$ of 9.4.2 and Lemma 9.5.1 we get a homeomorphism $\psi : |M| \to C_{f_1}$. One checks that $\{(M,\psi),(M_s',\varphi_s)\}$ is a triangulation of γ. Q.E.D.

9.6. Let $A \subset R^m$ and $B \subset R^n$ be polyhedra. A continuous map $f : A \to B$ is called piecewise linear if its graph is a polyhedron in $R^m \times R^n = R^{m+n}$.

If $s : K \to L$ is a s.m., then $|s| : |K| \to |L|$ is piecewise linear. The following partial converse is also true: given a proper piecewise linear map $f : A \to B$, there exist linear triangulations of A and B such that f is simplicial with respect to them (see [Hu], Theorem 3.6).

References:

[Br] G. Bredon: Introduction to Compact Transformation Groups. Academic Press. New York and London. 1972.

[Ca] S. S. Cairns: Triangulation of the manifold of class one. Bull. Amer. Math. Soc. 41 (1935), 549-552.

[C] J. Cerf: Topologie de certains espaces de plongements. Bull. Soc. Math. France 89 (1961), 227-380.

[Coh] M. Cohen: Simplicial structures and transverse cellularity. Ann. of Math. 85 (1967), 218-245.

[Cos] M. Coste: Ensembles semi-algébriques. In "Géométrie Algébrique Réelle et Formes Quadratiques". Lecture Notes in Math. 959. Springer-Verlag 1982, pp. 109-138.

[De] H. Delfs: Kohomologie affiner semi-algebraischer Raüme. Thesis. University of Regensburg, 1980.

[E-S] S. Eilenberg, N. Steenrod: Foundations of Algebraic Topology. Princeton University Press. Princeton. 1952.

[Gib] C. G. Gibson et al.: Topological Stability of Smooth Mappings. Lecture Notes in Math. 552. Springer-Verlag, 1976.

[Gi] B. Giesecke: Simpliziale Zerlegung abzählbarer analytischer Raüme. Math. Z. 83 (1964), 177-213.

[Go] R. Godement: Théorie des Faisceaux. Hermann. Paris, 1964.

[G_1] M. Goresky: Triangulation of stratified objects. Proc. Amer. Math. Soc. 72 (1978), 193-200.

[G_2] M. Goresky: Whitney stratified chains and cochains. Trans. Amer. Math. Soc. 261 (1981), 175-196.

[Ha_1] R. Hardt: Stratification of real analytic mappings and images. Inventiones Math. 28 (1975), 193-208.

[Ha_2] R. Hardt: Triangulation of subanalytic sets and proper light subanalytic maps. Inventiones Math. 38 (1977), 207-217.

[He] E. Hendricks: Triangulation of stratified sets. Thesis. M.I.T., 1973.

[Hi_1] H. Hironaka: Triangulation of algebraic sets. In "Algebraic Geometry. Arcata 1974". Proc. Sympos. Pure Math. 29. Providence, R.I., 1975, pp. 165-185.

[Hi$_2$] H. Hironaka: Stratification and flatness. In "Real and Complex Singularities". Sijthoff and Noordhoff. Oslo, 1977.

[Hu] J. Hudson: Piecewise Linear Topology. Benjamin. New York 1969.

[Il] S. Illman: Smooth equivariant triangulations of G-manifolds for G a finite group. Math. Ann. 233 (1978), 199-220.

[J$_1$] F.E.A. Johnson: On the triangulation of smooth fibre bundles. To appear in Fund. Math.

[J$_2$] F.E.A. Johnson: On the triangulation of stratified sets and singular varieties. Trans. Amer. Math. Soc. 275 (1983), 333-343.

[J$_3$] F.E.A. Johnson: On the presentation of stratified sets and singular varieties. Mathematika 29 (1982), 137-170.

[Ka] M. Kato: Elementary topology of analytic sets. Sugaku 25 (1973), 38-51 (in Japanese).

[K-B] B.C. Koopman, A. B. Brown: On the covering of analytic loci by complexes. Trans. Amer. Math. Soc. 34 (1932), 231-251.

[Le] S. Lefschetz: Topology. Amer. Math. Soc. Coll. Publications. New York, 1930.

[L-W] S. Lefschetz, J. H. C. Whitehead: On analytical complexes. Trans. Amer. Math. Soc. 35 (1933), 510-517.

[Lel] N.W. Lellman: Orbitraüme von G-Mannigfaltigkeiten und stratifizierte Mengen. Diplomarbeit, Bonn 1975.

[Lo] S. Lojasiewicz: Triangulation of semi-analytic sets. Ann. Scuola Norm. Sup. Pisa 18 (1964), 449-474.

[Ma$_1$] J.N. Mather: Notes on Topological Stability. Mimeographed Notes. Harvard, 1970.

[Ma$_2$] J.N. Mather: How to stratify mappings and jet spaces. In "Singularités d'applications Différentiables". Lecture Notes in Math. 535. Springer-Verlag 1976, pp. 128-176.

[Ma$_3$] J.N. Mather: Stratifications and mappings. In "Dynamical Systems". Academic Press, 1973, pp. 195-223.

[Mat] T. Matumoto: Equivariant stratification of a compact differentiable transformation group. Preprint. 1977.

[Mu] J. Munkres: Elementary Differential Topology. Annals of Math. Studies 54. Princeton University Press. Princeton, 1966.

[P] H. Putz: Triangulation of fibre bundles. Canadian J. of Math. 19 (1967), 499-513.

[T_1] R. Thom: La stabilité topologique des applications polynomiales. Enseignement Mathématique 8 (1962), 24-33.

[T_2] R. Thom: Ensembles et morphismes stratifiés. Bull. Amer. Math. Soc. 75 (1969), 240-284.

[T_3] R. Thom: Local topological properties of differentiable mappings. In "Differential Analysis". Oxford University Press, 1964, pp. 191-202.

[Ve_1] A. Verona: Le théorème de de Rham pour les préstratifications abstraites. C. R. Acad. Sc. Paris 273 (1971), 886-889.

[Ve_2] A. Verona: Homological properties of abstract stratifications. Rev. Roum. Math. Pures et Appl. 17 (1972), 1109-1121.

[Ve_3] A. Verona: Triangulation of stratified fibre bundles. Manuscripta math. 30 (1980), 425-445.

[Ve_4] A. Verona: Les applications topologiquement stables sont triangulables. C. R. Acad. Sc. Paris 296 (1983), 271-274.

[W] B.L. van der Waerden: Topologische Begründung des Kalküls der abzahlenden Geometrie. Math. Ann. 102 (1929), 337-362.

[Y] C.T. Yang: The triangulability of the orbit space of a differentiable transformation group. Bull. Amer. Math. Soc. 69 (1963), 405-408.

Subject index

Symbol Index

Vol. 952: Combinatorial Mathematics IX. Proceedings, 1981. Edited by E. Billington, S. Oates-Williams, and A.P. Street. XI, 443 pages. 1982.

Vol. 953: Iterative Solution of Nonlinear Systems of Equations. Proceedings, 1982. Edited by R. Ansorge, Th. Meis, and W. Törnig. VII, 202 pages. 1982.

Vol. 954: S.G. Pandit, S.G. Deo, Differential Systems Involving Impulses. VII, 102 pages. 1982.

Vol. 955: G. Gierz, Bundles of Topological Vector Spaces and Their Duality. IV, 296 pages. 1982.

Vol. 956: Group Actions and Vector Fields. Proceedings, 1981. Edited by J.B. Carrell. V, 144 pages. 1982.

Vol. 957: Differential Equations. Proceedings, 1981. Edited by D.G. de Figueiredo. VIII, 301 pages. 1982.

Vol. 958: F.R. Beyl, J. Tappe, Group Extensions, Representations, and the Schur Multiplicator. IV, 278 pages. 1982.

Vol. 959: Géométrie Algébrique Réelle et Formes Quadratiques, Proceedings, 1981. Edité par J.-L. Colliot-Thélène, M. Coste, L. Mahé, et M.-F. Roy. X, 458 pages. 1982.

Vol. 960: Multigrid Methods. Proceedings, 1981. Edited by W. Hackbusch and U. Trottenberg. VII, 652 pages. 1982.

Vol. 961: Algebraic Geometry. Proceedings, 1981. Edited by J.M. Aroca, R. Buchweitz, M. Giusti, and M. Merle. X, 500 pages. 1982.

Vol. 962: Category Theory. Proceedings, 1981. Edited by K.H. Kamps D. Pumplün, and W. Tholen, XV, 322 pages. 1982.

Vol. 963: R. Nottrot, Optimal Processes on Manifolds. VI, 124 pages. 1982.

Vol. 964: Ordinary and Partial Differential Equations. Proceedings, 1982. Edited by W.N. Everitt and B.D. Sleeman. XVIII, 726 pages. 1982.

Vol. 965: Topics in Numerical Analysis. Proceedings, 1981. Edited by P.R. Turner. IX, 202 pages. 1982.

Vol. 966: Algebraic K-Theory. Proceedings, 1980, Part I. Edited by R.K. Dennis. VIII, 407 pages. 1982.

Vol. 967: Algebraic K-Theory. Proceedings, 1980. Part II. VIII, 409 pages. 1982.

Vol. 968: Numerical Integration of Differential Equations and Large Linear Systems. Proceedings, 1980. Edited by J. Hinze. VI, 412 pages. 1982.

Vol. 969: Combinatorial Theory. Proceedings, 1982. Edited by D. Jungnickel and K. Vedder. V, 326 pages. 1982.

Vol. 970: Twistor Geometry and Non-Linear Systems. Proceedings, 1980. Edited by H.-D. Doebner and T.D. Palev. V, 216 pages. 1982.

Vol. 971: Kleinian Groups and Related Topics. Proceedings, 1981. Edited by D.M. Gallo and R.M. Porter. V, 117 pages. 1983.

Vol. 972: Nonlinear Filtering and Stochastic Control. Proceedings, 1981. Edited by S.K. Mitter and A. Moro. VIII, 297 pages. 1983.

Vol. 973: Matrix Pencils. Proceedings, 1982. Edited by B. Kågström and A. Ruhe. XI, 293 pages. 1983.

Vol. 974: A. Draux, Polynômes Orthogonaux Formels – Applications. VI, 625 pages. 1983.

Vol. 975: Radical Banach Algebras and Automatic Continuity. Proceedings, 1981. Edited by J.M. Bachar, W.G. Bade, P.C. Curtis Jr., H.G. Dales and M.P. Thomas. VIII, 470 pages. 1983.

Vol. 976: X. Fernique, P.W. Millar, D.W. Stroock, M. Weber, Ecole d'Eté de Probabilités de Saint-Flour XI – 1981. Edited by P.L. Hennequin. XI, 465 pages. 1983.

Vol. 977: T. Parthasarathy, On Global Univalence Theorems. VIII, 106 pages. 1983.

Vol. 978: J. Ławrynowicz, J. Krzyż, Quasiconformal Mappings in the Plane. VI, 177 pages. 1983.

Vol. 979: Mathematical Theories of Optimization. Proceedings, 1981. Edited by J.P. Cecconi and T. Zolezzi. V, 268 pages. 1983.

Vol. 980: L. Breen. Fonctions thêta et théorème du cube. XIII, 115 pages. 1983.

Vol. 981: Value Distribution Theory. Proceedings, 1981. Edited by I. Laine and S. Rickman. VIII, 245 pages. 1983.

Vol. 982: Stability Problems for Stochastic Models. Proceedings, 1982. Edited by V.V. Kalashnikov and V.M. Zolotarev. XVII, 295 pages. 1983.

Vol. 983: Nonstandard Analysis-Recent Developments. Edited by A.E. Hurd. V 213 pages. 1983.

Vol. 984: A. Bove, J.E. Lewis, C. Parenti, Propagation of Singularities for Fuchsian Operators. IV, 161 pages. 1983.

Vol. 985: Asymptotic Analysis II. Edited by F. Verhulst. VI, 497 pages. 1983.

Vol. 986: Séminaire de Probabilités XVII 1981/82. Proceedings. Edited by J. Azéma and M. Yor. V, 512 pages. 1983.

Vol. 987: C.J. Bushnell, A. Fröhlich, Gauss Sums and p-adic Division Algebras. XI, 187 pages. 1983.

Vol. 988: J. Schwermer, Kohomologie arithmetisch definierter Gruppen und Eisensteinreihen. III, 170 pages. 1983.

Vol. 989: A.B. Mingarelli, Volterra-Stieltjes Integral Equations and Generalized Ordinary Differential Expressions. XIV, 318 pages. 1983.

Vol. 990: Probability in Banach Spaces IV. Proceedings, 1982. Edited by A. Beck and K. Jacobs. V, 234 pages. 1983.

Vol. 991: Banach Space Theory and its Applications. Proceedings, 1981. Editec by A. Pietsch, N. Popa and I. Singer. X, 302 pages. 1983.

Vol. 992: Harmonic Analysis, Proceedings, 1982. Edited by G. Mauceri, F. Ricci and G. Weiss. X, 449 pages. 1983.

Vol. 993: R.D. Bourgin, Geometric Aspects of Convex Sets with the Radon-Nikodým Property. XII, 474 pages. 1983.

Vol. 994: J.-L. Journé, Calderón-Zygmund Operators, Pseudo-Differential Operators and the Cauchy Integral of Calderón. VI, 129 pages. 1983.

Vol. 995: Banach Spaces, Harmonic Analysis, and Probability Theory. Proceedings, 1980–1981. Edited by R.C. Blei and S.J. Sidney. V, 173 pages. 1983.

Vol. 996: Invariant Theory. Proceedings, 1982. Edited by F. Gherardelli. V, 159 pages. 1983.

Vol. 997: Algebraic Geometry – Open Problems. Edited by C. Ciliberto, F. Ghione and F. Orecchia. VIII, 411 pages. 1983.

Vol. 998: Recent Developments in the Algebraic, Analytical, and Topological Theory of Semigroups. Proceedings, 1981. Edited by K.H. Hofmann, H. Jürgensen and H.J. Weinert. VI, 486 pages. 1983.

Vol. 999: C. Preston, Iterates of Maps on an Interval. VII, 205 pages. 1983.

Vol. 1000: H. Hopf, Differential Geometry in the Large, VII, 184 pages. 1983.

Vol. 1001: D.A. Hejhal, The Selberg Trace Formula for PSL(2, $\mathbb{R}$). Volume 2. VIII, 806 pages. 1983.

Vol. 1002: A. Edrei, E.B. Saff, R.S. Varga, Zeros of Sections of Power Series. VIII, 115 pages. 1983.

Vol. 1003: J. Schmets, Spaces of Vector-Valued Continuous Functions. VI, 117 pages. 1983.

Vol. 1004: Universal Algebra and Lattice Theory. Proceedings, 1982. Edited by R.S. Freese and O.C. Garcia. VI, 308 pages. 1983.

Vol. 1005: Numerical Methods. Proceedings, 1982. Edited by V. Pereyra and A. Reinoza. V, 296 pages. 1983.

Vol. 1006: Abelian Group Theory. Proceedings, 1982/83. Edited by R. Göbel, L. Lady and A. Mader. XVI, 771 pages. 1983.

Vol. 1007: Geometric Dynamics. Proceedings, 1981. Edited by J. Palis Jr. IX, 827 pages. 1983.

Vol. 1008: Algebraic Geometry. Proceedings, 1981. Edited by J. Dolgachev. V, 138 pages. 1983.

Vol. 1009: T. A. Chapman, Controlled Simple Homotopy Theory and Applications. III, 94 pages. 1983.

Vol. 1010: J.-E. Dies, Chaînes de Markov sur les permutations. IX, 226 pages. 1983.

Vol. 1011: J. M. Sigal. Scattering Theory for Many-Body Quantum Mechanical Systems. IV, 132 pages. 1983.

Vol. 1012: S. Kantorovitz, Spectral Theory of Banach Space Operators. V, 179 pages. 1983.

Vol. 1013: Complex Analysis – Fifth Romanian-Finnish Seminar. Part 1. Proceedings, 1981. Edited by C. Andreian Cazacu, N. Boboc, M. Jurchescu and I. Suciu. XX, 393 pages. 1983.

Vol. 1014: Complex Analysis – Fifth Romanian-Finnish Seminar. Part 2. Proceedings, 1981. Edited by C. Andreian Cazacu, N. Boboc, M. Jurchescu and I. Suciu. XX, 334 pages. 1983.

Vol. 1015: Equations différentielles et systèmes de Pfaff dans le champ complexe – II. Seminar. Edited by R. Gérard et J. P. Ramis. V, 411 pages. 1983.

Vol. 1016: Algebraic Geometry. Proceedings, 1982. Edited by M. Raynaud and T. Shioda. VIII, 528 pages. 1983.

Vol. 1017: Equadiff 82. Proceedings, 1982. Edited by H. W. Knobloch and K. Schmitt. XXIII, 666 pages. 1983.

Vol. 1018: Graph Theory, Łagów 1981. Proceedings, 1981. Edited by M. Borowiecki, J. W. Kennedy and M. M. Sysło. X, 289 pages. 1983.

Vol. 1019: Cabal Seminar 79–81. Proceedings, 1979–81. Edited by A. S. Kechris, D. A. Martin and Y. N. Moschovakis. V, 284 pages. 1983.

Vol. 1020: Non Commutative Harmonic Analysis and Lie Groups. Proceedings, 1982. Edited by J. Carmona and M. Vergne. V, 187 pages. 1983.

Vol. 1021: Probability Theory and Mathematical Statistics. Proceedings, 1982. Edited by K. Itô and J.V. Prokhorov. VIII, 747 pages. 1983.

Vol. 1022: G. Gentili, S. Salamon and J.-P. Vigué. Geometry Seminar "Luigi Bianchi", 1982. Edited by E. Vesentini. VI, 177 pages. 1983.

Vol. 1023: S. McAdam, Asymptotic Prime Divisors. IX, 118 pages. 1983.

Vol. 1024: Lie Group Representations I. Proceedings, 1982–1983. Edited by R. Herb, R. Lipsman and J. Rosenberg. IX, 369 pages. 1983.

Vol. 1025: D. Tanré, Homotopie Rationnelle: Modèles de Chen, Quillen, Sullivan. X, 211 pages. 1983.

Vol. 1026: W. Plesken, Group Rings of Finite Groups Over p-adic Integers. V, 151 pages. 1983.

Vol. 1027: M. Hasumi, Hardy Classes on Infinitely Connected Riemann Surfaces. XII, 280 pages. 1983.

Vol. 1028: Séminaire d'Analyse P. Lelong – P. Dolbeault – H. Skoda. Années 1981/1983. Edité par P. Lelong, P. Dolbeault et H. Skoda. VIII, 328 pages. 1983.

Vol. 1029: Séminaire d'Algèbre Paul Dubreil et Marie-Paule Malliavin. Proceedings, 1982. Edité par M.-P. Malliavin. V, 339 pages. 1983.

Vol. 1030: U. Christian, Selberg's Zeta-, L-, and Eisensteinseries. XII, 196 pages. 1983.

Vol. 1031: Dynamics and Processes. Proceedings, 1981. Edited by Ph. Blanchard and L. Streit. IX, 213 pages. 1983.

Vol. 1032: Ordinary Differential Equations and Operators. Proceedings, 1982. Edited by W. N. Everitt and R. T. Lewis. XV, 521 pages. 1983.

Vol. 1033: Measure Theory and its Applications. Proceedings, 1982. Edited by J. M. Belley, J. Dubois and P. Morales. XV, 317 pages. 1983.

Vol. 1034: J. Musielak, Orlicz Spaces and Modular Spaces. V, 222 pages. 1983.

Vol. 1035: The Mathematics and Physics of Disordered Media. Proceedings, 1983. Edited by B. D. Hughes and B. W. Ninham. VII, 432 pages. 1983.

Vol. 1036: Combinatorial Mathematics X. Proceedings, 1982. Edited by L. R. A. Casse. XI, 419 pages. 1983.

Vol. 1037: Non-linear Partial Differential Operators and Quantization Procedures. Proceedings, 1981. Edited by S. I. Andersson and H.-D. Doebner. VII, 334 pages. 1983.

Vol. 1038: F. Borceux, G. Van den Bossche, Algebra in a Localic Topos with Applications to Ring Theory. IX, 240 pages. 1983.

Vol. 1039: Analytic Functions, Błażejewko 1982. Proceedings. Edited by J. Ławrynowicz. X, 494 pages. 1983

Vol. 1040: A. Good, Local Analysis of Selberg's Trace Formula. III, 128 pages. 1983.

Vol. 1041: Lie Group Representations II. Proceedings 1982–1983. Edited by R. Herb, S. Kudla, R. Lipsman and J. Rosenberg. IX, 340 pages. 1984.

Vol. 1042: A. Gut, K. D. Schmidt, Amarts and Set Function Processes. III, 258 pages. 1983.

Vol. 1043: Linear and Complex Analysis Problem Book. Edited by V. P. Havin, S. V. Hruščëv and N. K. Nikol'skii. XVIII, 721 pages. 1984.

Vol. 1044: E. Gekeler, Discretization Methods for Stable Initial Value Problems. VIII, 201 pages. 1984.

Vol. 1045: Differential Geometry. Proceedings, 1982. Edited by A. M. Naveira. VIII, 194 pages. 1984.

Vol. 1046: Algebraic K-Theory, Number Theory, Geometry and Analysis. Proceedings, 1982. Edited by A. Bak. IX, 464 pages. 1984.

Vol. 1047: Fluid Dynamics. Seminar, 1982. Edited by H. Beirão da Veiga. VII, 193 pages. 1984.

Vol. 1048: Kinetic Theories and the Boltzmann Equation. Seminar, 1981. Edited by C. Cercignani. VII, 248 pages. 1984.

Vol. 1049: B. Iochum, Cônes autopolaires et algèbres de Jordan. VI, 247 pages. 1984.

Vol. 1050: A. Prestel, P. Roquette, Formally p-adic Fields. V, 167 pages. 1984.

Vol. 1051: Algebraic Topology, Aarhus 1982. Proceedings. Edited by I. Madsen and B. Oliver. X, 665 pages. 1984.

Vol. 1052: Number Theory. Seminar, 1982. Edited by D.V. Chudnovsky, G.V. Chudnovsky, H. Cohn and M.B. Nathanson. V, 309 pages. 1984.

Vol. 1053: P. Hilton, Nilpotente Gruppen und nilpotente Räume. V, 221 pages. 1984.

Vol. 1054: V. Thomée, Galerkin Finite Element Methods for Parabolic Problems. VII, 237 pages. 1984.

Vol. 1055: Quantum Probability and Applications to the Quantum Theory of Irreversible Processes. Proceedings, 1982. Edited by L. Accardi, A. Frigerio and V. Gorini. VI, 411 pages. 1984.

Vol. 1056: Algebraic Geometry. Bucharest 1982. Proceedings, 1982. Edited by L. Bădescu and D. Popescu. VII, 380 pages. 1984.

Vol. 1057: Bifurcation Theory and Applications. Seminar, 1983. Edited by L. Salvadori. VII, 233 pages. 1984.

Vol. 1058: B. Aulbach, Continuous and Discrete Dynamics near Manifolds of Equilibria. IX, 142 pages. 1984.

Vol. 1059: Séminaire de Probabilités XVIII, 1982/83. Proceedings. Edité par J. Azéma et M. Yor. IV, 518 pages. 1984.

Vol. 1060: Topology. Proceedings, 1982. Edited by L. D. Faddeev and A. A. Mal'cev. VI, 389 pages. 1984.

Vol. 1061: Séminaire de Théorie du Potentiel. Paris, No. 7. Proceedings. Directeurs: M. Brelot, G. Choquet et J. Deny. Rédacteurs: F. Hirsch et G. Mokobodzki. IV, 281 pages. 1984.